U0318524

METAMORPHOSIS

Derivation and Breakthrough

Basic Form

变形记

建筑立面的衍生与突破 上册

基本形

香港理工国际出版社 编著

http://www.hustp.com

ART COMES FROM REALITY BUT

艺术源自于现实而终止于抽象，

ENDS IN ABSTRACT,

而建筑作品则相反，

WHILE ARCHITECTURAL WORKS ARE

源自于抽象概念，形成于现实。

ON THE CONTRARY.

By US Modernism Architect John Hejduk

美国著名现代主义建筑师约翰·海杜克

IN WALKING TO NEW BUILDINGS,

勒·柯布西耶在

LECORBUSIER SAYS,

《走向新建筑》中说过：

"ARCHITECTURE HAS NOTHING TO DO WITH

"建筑跟各种风格无关，

STYLES AND IT STIMULATES ITS POTENTIAL

建筑以它的抽象性激发最高级的才能。

THROUGH THE ABSTRACT FEATURE WHICH IS

建筑的抽象性

SO UNIQUE AND GLORIOUS

具有如此独特又如此辉煌的能力，

THAT IT IS DEEPLY ROOTED IN THE

以至假如它扎根在俗物中，

EARTHLINGS. AND VOLUME AND SURFACE

它能把俗物精神化……

ARE THE KEY ELEMENTS TO REPRESENT

体块和表面

ARCHITECTURE."

是建筑借以表现自己的要素。"

序言

In recent years, with technology development and blend of international design concepts, the architectural surface makes breakthroughs in the original shapes and aesthetic framework and focuses more on humanity and environment to provide creative sources for the surface design. Now the surface design revives with the information wave, and becomes the primary means for architects to represent the building image. It is now increasingly important.

From the contextual perspective, the plane contains the plane itself and its composition. The plane is the two dimensional feature while the composition refers to the composition of structure. Another complete definition is to disintegrate all existing forms into a perfect form. The plane composition creates a rational, orderly and abstract visual beauty through the unique visual form and shape.

The surface composition is represented in the form to bring rich effects. As the architectural surface becomes increasingly diverse, creation in surface has become the pursuit of a designer. The book tries to analyze the surface and dig out elements and compositions and the mental experience in accordance with the high definition details, offering the complete surface design concepts.

In the surface demonstration, geometric transition is mostly used, to make the abstract the architectural form become the basic form. The form beauty of basic forms gives two different mental differences, namely the different mental feeling of shape, structure and form, and the architectural concept of history and culture which contains the accepted mental feeling. Thus the book is divided into three categories of basic, creative and bionics forms and it is different from most books published. The book summarizes the plane form and composition and comes up with the surface design strategy which is illustrated by cases to take on the diverse development trend of architectural surfaces that can be drawn upon by the designer.

Architects abroad have obtained rich experiences in surface design. Development of new materials and technology has brought opportunities to the surface diversity. On the contrary, domestic architectural design started late and it is still in the stage of passive absorption and digestion. Meanwhile domestic design blindly pursues the form creation and ignores the relationship between architectural concept and local culture. Anyway, the domestic design represents the ground-breaking trial in architecture image. For the project selection, the book selects primarily the foreign projects and some excellent domestic projects and we hope the successful cases can inspire domestic designers more.

近年来，随着科技的进步和国际间设计理念的交融，建筑表皮的设计方法打破了以往单纯的造型表现和美学追求的框架，而是更多地关注人文主义和环境，并通过寻求技术与艺术的融合，为建筑表皮的创新设计提供源泉。建筑表皮设计在如今的信息化浪潮中觉醒过来，成为建筑师们表达建筑形象最常用的表现手段。建筑表皮逐渐占据着日趋重要的地位。

平面构成从字义上理解，可以分为两个方面来认识——“平面”和“构成”，平面是指其所运用的形体所表现出来的二维平面特征；“构成”则是指形体的组合方式。另一种较完整的定义是：将既有的形态（包括具象形态和抽象形态——点、线、面）在二维平面内按照一定的秩序和法则进行分解、组合，从而构成理想形态的组合形式。平面构成以其特有的视觉形态和构成形式组成严谨但又不乏节奏感的画面，营造一种理性、秩序与抽象的视觉美感。

建筑表皮的平面构成表现形式主要是利用平面构成原理以不同的组合方式在建筑表皮上大作构成文章，创造出丰富多彩的构成效果。在建筑表皮的表现形式日益丰富和多元化的今天，表皮的创新无疑成了设计师极力追求的目标。本书试图从平面构成的独特角度对建筑表皮进行深入分析，挖掘平面构成各要素及其构成方式以及其引发的心理体验，结合高清细部全彩图，给设计师呈现最完整、立体的创意表皮设计构思。

在建筑表皮的平面表现形式中，最通常的做法就是几何转换，即把建筑形态抽象为基本几何形，来创造独特的建筑形象。基本几何形的形态美给人带来的视觉心理差异可分为两种：一种是由于形态本身的不同外形、结构、形式给人造成的不同心理感受；另一种则是由于历史、文化的沉淀令人自然而然地想到建筑的理念与个性，它蕴涵着人们约定俗成的心理感受。本书由此划分基本形、创意形、仿生形三大类别，区别于市面上的大多数同类图书，从平面构成的角度，通过总结归纳其构成形式和方法，得出建筑表皮的设计策略，并通过成功案例进行说明论证，呈现建筑表皮的多元化发展趋势，便于设计师参考与借鉴。

在国外，建筑师们在表皮设计方面已积累了丰富的经验，新材料和新技术日新月异，为建筑表皮的多样化发展创造了契机。相对于国外，国内的建筑表皮设计起步较晚，还处于吸收和消化阶段，在技术和表现形式方面依然相对落后；同时盲目地追求外形上的标新立异，忽视建筑理念与地域文化的关联。但无论其功过得失，都代表了国内设计师对建筑形象创作的突破性尝试。而本书对于项目的筛选，主要侧重于以国外项目为主，再精选出一些国内的优秀项目汇集成册，希望能通过这些成功的案例带给国内设计师更多的思考与启发。

BASIC FORM

基本形

CONTENTS
目录

BASIC FORM
基本形

基本形

THE US ARCHITECTURE CRITIC
美国著名建筑历史理论家
FRANCIS DERCHIN SAYS,
弗朗西斯.德钦曾说：
"THE FORM, PURER AND MORE
“形态越是规矩和单纯，
CUSTOMARY, IS EASIER TO
越容易被感知和理解。”
PERCEIVE AND UNDERSTAND."

PURE, NEAT, STRIKING

纯粹、规整、形象鲜明

Colors can be reduced into red, yellow and blue while diverse architectural shapes are composed of basic forms. With the development of world economies and increase in international exchanges, architecture is steadily making progress as can be seen in the emergence of residential, sports, commerce, cultural and tourist facilities of unique form. The buildings are characterized by novel façade, complex planar combinations, basic geometrical forms, all of which together take on the various facades. For example, the circle, oval, triangle and curved forms are used to form the simple façade, easy to walk close to.

The basic planar forms like rectangle, trapezoid, parallelogram, circle, triangle, etc make up the architectural body, striking and clear. Thus, those forms are aesthetically recognized and bring to us spiritual enjoyment. Besides, they exist only confined by themselves. Under such circumstances, the architectural function, material and volume only consider to offer a pleasant visual experience.

Currently, it has become a design trend to maximally economize on materials, purify the surface, discard all non-essential elements, and interconnect life with space in order to present a multi-level façade and to enrich the whole architectural space through the basic, pure geometric forms. Specifically, the basic geometric shapes can create the most harmonious architecture forms and their combinations can bring cordial and lifelike feelings.

In addition, the basic planar forms can be the life of architectural design and the designer captures the "picture" inside to freely present the space ideology and to maximally display individuality and creativity. For instance, square, circle, rhombus, linear or any other planar forms, when processed in a abstract, diverse way, can make buildings look rich in form and unique in beauty. Then the spaces won't be of matter and are offered a spiritual meaning apart from distance and volunme.

Square / Linear / Triangular / Rhombus / Round / Mixed

方形 / 线形 / 三角形 / 菱形 / 圆形 / 混合形

五彩缤纷的颜色可以分解为红、黄、蓝三原色，丰富多姿的建筑形体也可以由最基本的几何形态构成。当前随着各国经济的发展以及国际交流日益增多，建筑技术在不断地进步，各种外观造型新颖的住宅、体育、商业、文化、旅游等建筑不断出现。这些建筑的特点是：立面造型新颖、平面组合复杂、利用各种最基本的几何图形组合成建筑的立面和平面，构成了丰富多彩的外立面形象，例如圆弧形、椭圆形、三角形、曲线形等平面图形的运用，它们构成了简洁而不简单的建筑外观，让人感到平易近人。

这些基本的平面状态是最纯粹的基本形式，由矩形、梯形、平行四边形、圆形、三角形等构成的建筑形体，形象鲜明、实在、毫不含糊，由于这个原因，这些形式一直得到美的认可，并带给人们精神上的享受；这些形体，作为建筑平面在几何状态下组合的母体，以它们为原型进行组织，可以得到多种平面形态，而它们的存在状态只与自身的约束条件有关，在这样的状态下，功能、材料、尺度等在实际操作中可以不再考虑，只从自身条件出发带给人们规整又不失韵律的视觉体验。

在建筑设计的趋势下，要最大限度地简约材料，纯化表面，去除一切非本质的因素，把生命与空间交织在一起，以极端简化的形式，最大程度地创造平衡，可以通过这些最基本、单纯的几何形式，将多层次的复杂的建筑立面展现出来，丰富整个建筑空间。最基本的几何形态能够创造出和谐与丰富的建筑形体，它们之间的组合能够给人以亲切感与生命感。

另外，最基本的平面形态，可以成为建筑设计的生命，设计师把整个建筑作为一个画面空间的"场"来捕捉图形，使自由的空间意识充分展现出来，最大限度地体现着自身的个性和创造意识。例如方形、圆形、菱形、线形或者是任意的平面形式，利用这些基本的几何构图加上对空间作出抽象、多样性的想像，使建筑散发出丰富而独特的美，这时空间不再是物质的，它排除了距离、尺度等观念，而被赋予一种精神空间的内涵。

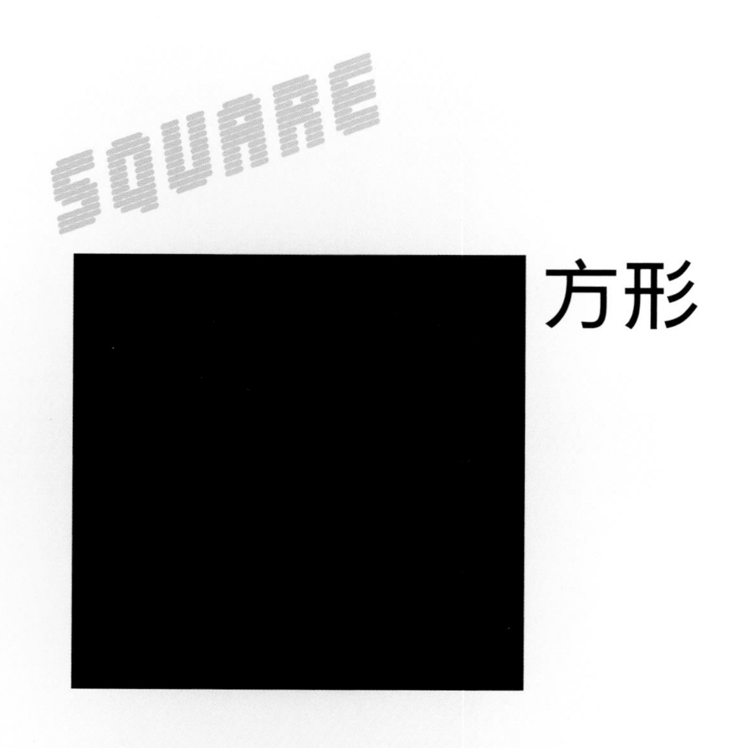

CAREFUL, NEAT, STEADY AND ORDERLY

严谨、整齐而不失平稳，秩序的抽象美

The façade of square buildings, when expressing certain thoughts or feelings like cultural and aesthetic values and modern senses inside, can't do it as exactly as paintings and sculptures and can only in the abstract way compare architectural art to abstract art language to connect with people in the symbol aesthetic language.

Square symbols or symbolic techniques can express the inner meanings inside architecture and thus extract symbolic figures to reach the aesthetic effect from "image" to "form". And abstract art pursues to express the simple contents and the basic forms derived from squares well represent the simple concept and reserved metaphor.

Squares are the basic geometric forms of many urban building. The form often offers powerful, solid resonance and it gives a exact, grand, austere, powerful, mechanical, orderly and steady feeling. Squares can make you feel secure and meanwhile acts as the symbol of bathing your emotion. When a building composed of several squares stand firmly in front of you, it offers shock and untold feelings. It is often the symbol of the designer's idea, concept and spiritual appeal.

In the various abstract geometric figures, squares display the stillness of buildings for its exact, orderly, steady individuality. When it comes to application, squares gradually transform the spiritual function to the modern matter function and out of requirements for function, structure and energy-saving, modern buildings use square forms. The general trend for modern buildings is the geometric abstract and standardized geometry shapes adjust to the commercial, social production means. In the anti-tradition wave, squares reflect the revolutionary spirit.

方形的建筑外立面，在表达某种构思或情感时，诸如建筑形象的文化价值、审美价值及所体现的时代感，无法像绘画、雕塑那样具体、写实，只能用抽象的方法将建筑艺术比喻成抽象化的艺术，用抽象的美学语言或象征性的美学语言与人们进行审美联系。

方形符号或建筑造型的象征性手法能够表达建筑体某种文化上的含义，进而提炼象征符号的形式，达到观念艺术由“象”到“意”的审美效果。抽象艺术追求以简洁的形、体表达丰富的内涵，而运用方形及其演变出来的有规则和容易识别的基本形状，则更好地体现了这种简洁的意念，体现出含蓄的隐喻作用。

方形是很多都市建筑的基础几何图形。这个形状常常让人产生坚实、有力的共鸣。它常给人坚实、明确、庄严、权威、强力、机械、秩序、稳重的感觉。方形能够令人有安全感，是能够寄托心情的象征。当一个由众多方形组成的建筑物矗立面前的时候，带给人们的是震撼和不可言喻的情愫渊源。它往往象征和诉说着设计者的思想和理念以及建筑所要呈现的精神诉求。

在不同的抽象几何图形中，方形构图以其严谨、敦厚、整齐、平稳的个性显示了建筑不可更动性和静态的平衡，充分表现了简约而不简单的含义。方形的构图在建筑中的运用逐渐由古代的精神功能转到了现代的物质功能方面，现代建筑使用方形平面和空间多出于功能、结构和节能等方面的要求。现代建筑的基本倾向是几何抽象性，标准化的几何形适应了迅速发展起来的工业化、社会化的生产方式和大众对“量”的需求。在反传统的浪潮中，方形的几何抽象性以时代的革命精神突出表现了时代特征。

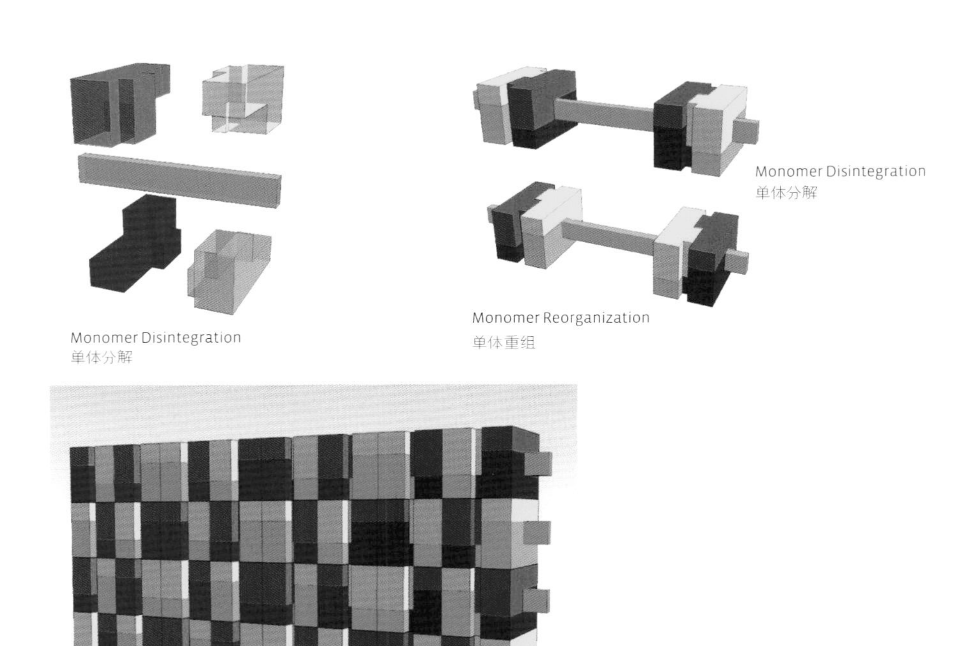

Monomer Disintegration 单体分解

Monomer Reorganization 单体重组

Monomer Disintegration 单体分解

Monomer Derivation 单体衍生

To design and shape a symbol place of locality, culture and uniqueness and carry the latest architectural space and life form, the designer integrates mountain, nature and living into the main building, creates a unique surface in the romantic way, and uses square combinations of different colors to make the building look like a colorful stone growing out of the ground.

为了设计并塑造一个具有本土性、文化性和独特性的标志性场所，以承载最新的建筑空间和生活状态，设计师将山体、自然、生长等要素融入到了主楼设计中，并以浪漫主义手法营造一种独特的建筑表皮形象，利用不同色块的方形平面组合，使建筑表面犹如一颗从大地中生长出的多彩宝石。

龙湖MOCO
龙湖MOCO
家居生活馆

Simple, striking white façade is mainly composed of two square planes: glass square and stone square. The surface of glass square is the massive ceiling-to-floor window to bring in natural light and landscape maximally. And the stone square faces the living room or study to create a peaceful atmosphere. The two square forms appear alternately and with the undulating white aluminum ornaments bring a multi-layered façade.

简洁、新锐的白色外表面，布局主要由两种方形平面组成：玻璃方形和石头方形框。玻璃方形表皮是大面积的落地玻璃，将自然光和景观最大化地引入进来；石头方形框对应的是卧室或者书房空间，营造安静、私密的环境氛围。这两种方形平面的交替出现外加立面上跳跃随机出现的白色铝板装饰体，形成多层次、富有节奏感的建筑外表皮形象。

The flexible use of squares is in fact an abstract process and it breaks the still façade and makes it moving. The façade arranges the squares in the L shape and thus opens the vertical façade to give the main building curved beauty and to enrich people's visual experience.

方形的加减法运用，本身就是一个抽象思维的过程，它打破了安静的建筑表皮，使立面充满灵动性。在建筑立面上将方框形作"L型"错落拼接排列，从而打开垂直立面，使主体建筑有了曲线的美感，丰富了人们的视觉体验。

The building looks as if made up of several brick or glass square boxes. The concept is rooted in the work area in a bank, namely 6×6 square. The 6-m cube is irregularly distributed on the façade and forms the open built-in open passage. Based upon such an area, the façade designs a sophisticated surface blended with city texture.

该大楼看上去像由许多砖或玻璃的方盒子构成。设计方案的理念基于银行理想的小型工作组面积，即6米乘以6米的正方形。这些边长6米的立方体不规则地排布在立面上，形成了内嵌式的开放通道。立面也根据这个面积设计出了像素般的外观，与城市文脉相融合。

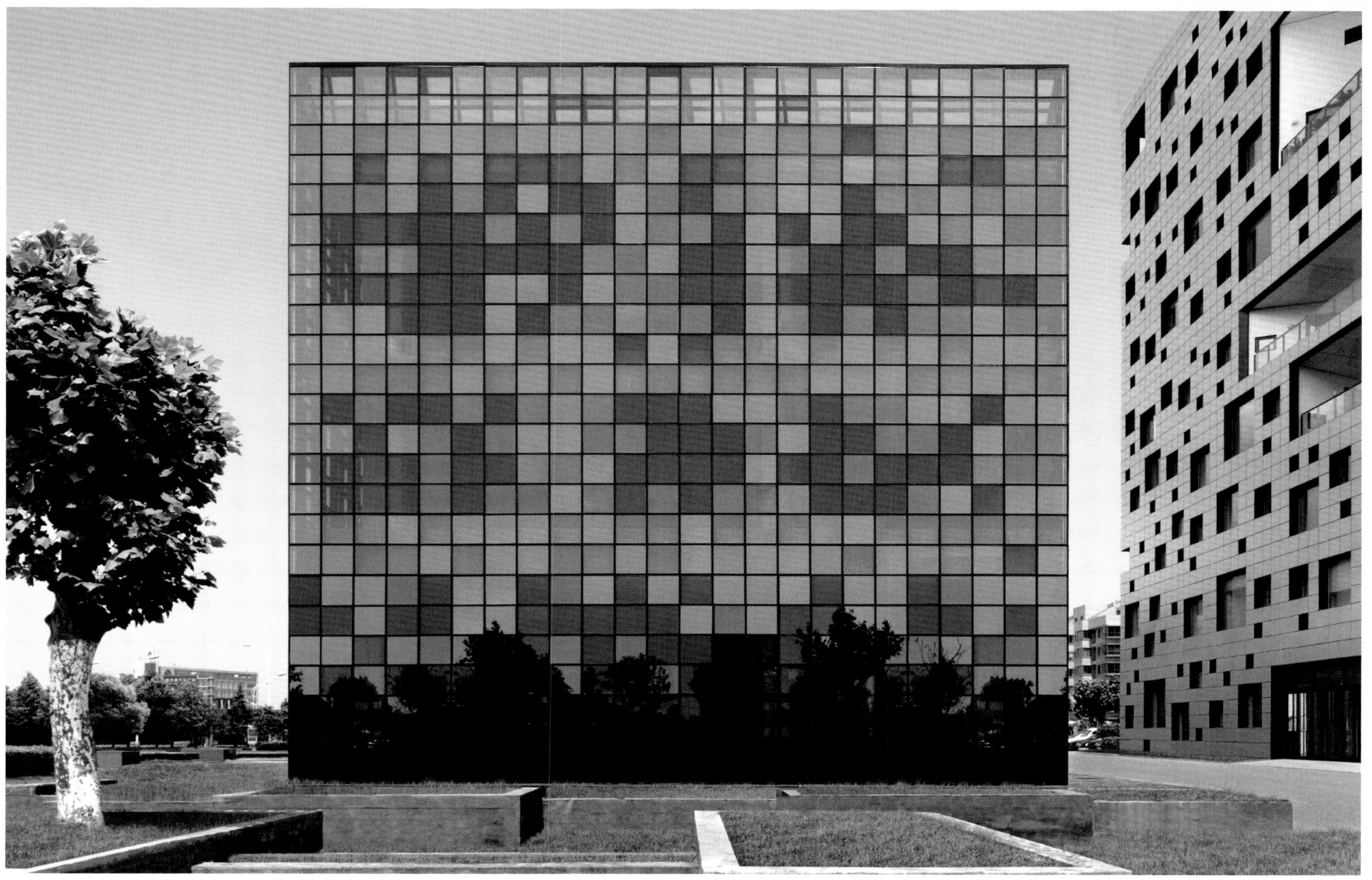

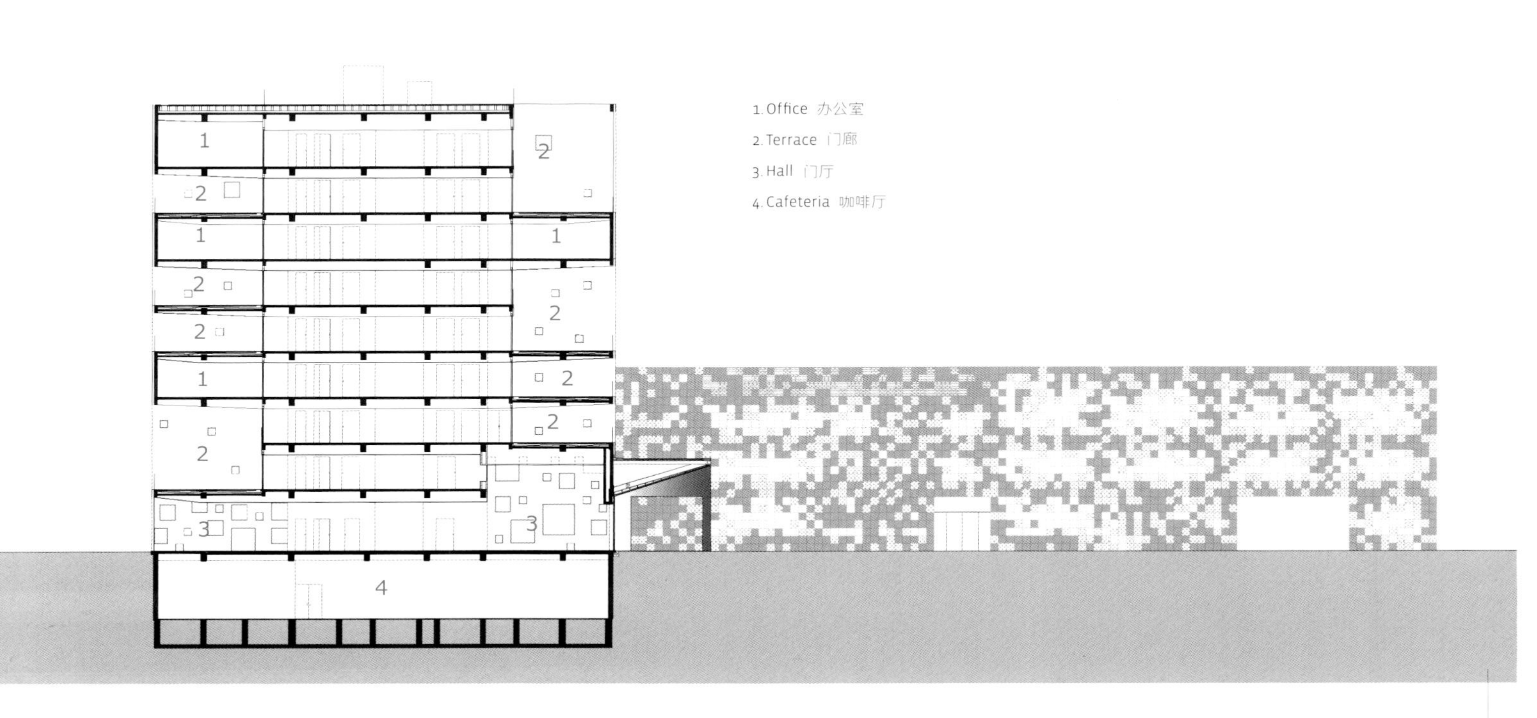

1. Office 办公室
2. Terrace 门廊
3. Hall 门厅
4. Cafeteria 咖啡厅

Here, you can obviously feel the forms of abstract and construction. The buildings join through squares of different sizes and form the leaping aesthetic and visual shock. The mosaic squares highlight the building volume. The rectangle buildings are fully interpreted here and blend with the surroundings.

在这里，能让人明显感受到抽象主义与构成主义的形式。这样的建筑通过大小不一的正方形与长方形的拼接来打开横、竖立面，在建筑样式上形成跳跃的美感，通过方形的凹凸感给人强烈的视觉刺激，从而给人留下深刻的印象；“马赛克”效果的方形平面，以深浅颜色的玻璃搭配来突显建筑体量。矩形建筑在这里得到了充分的演绎，玻璃的方形体块将建筑以更加亲和的姿态融入周边环境。

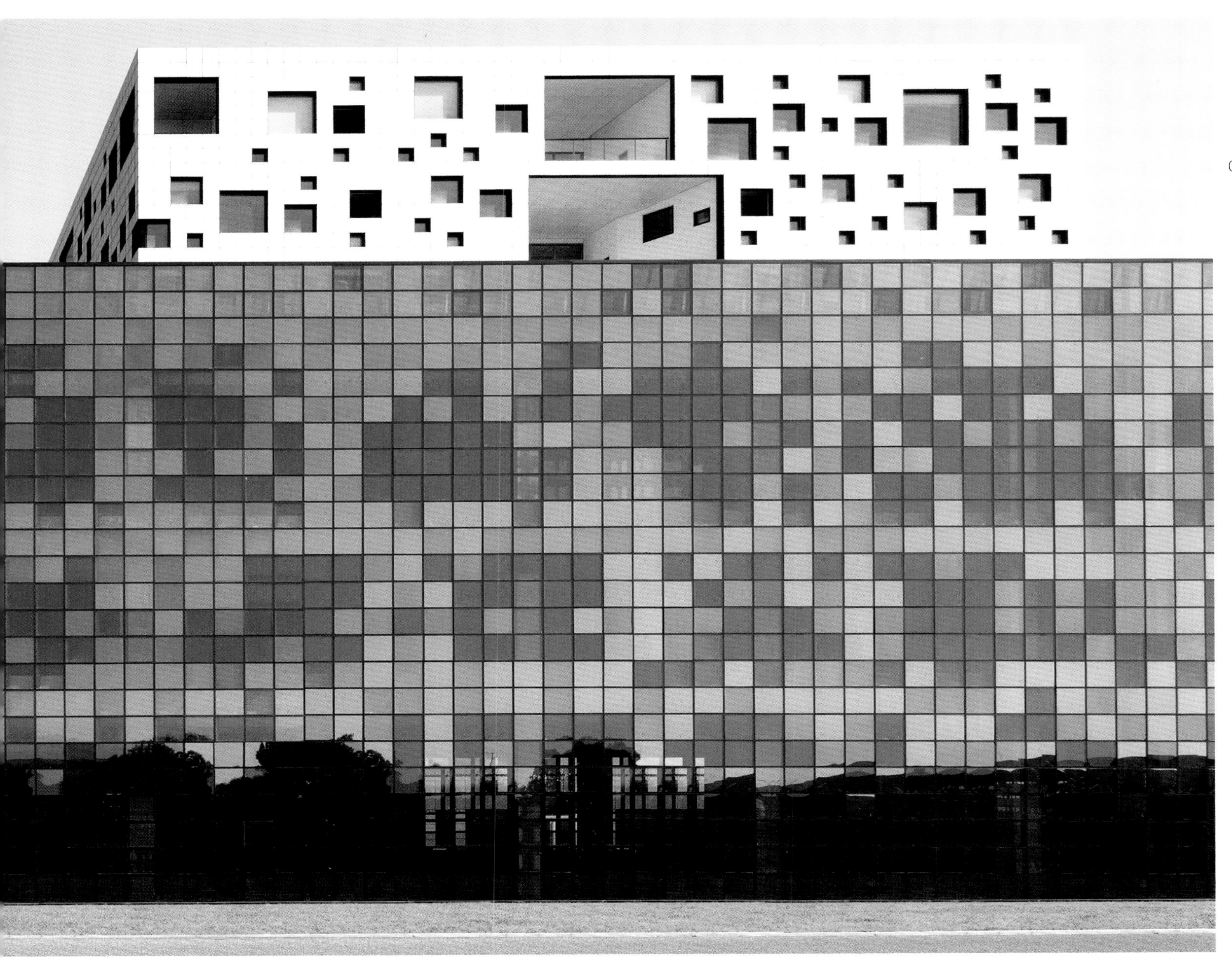

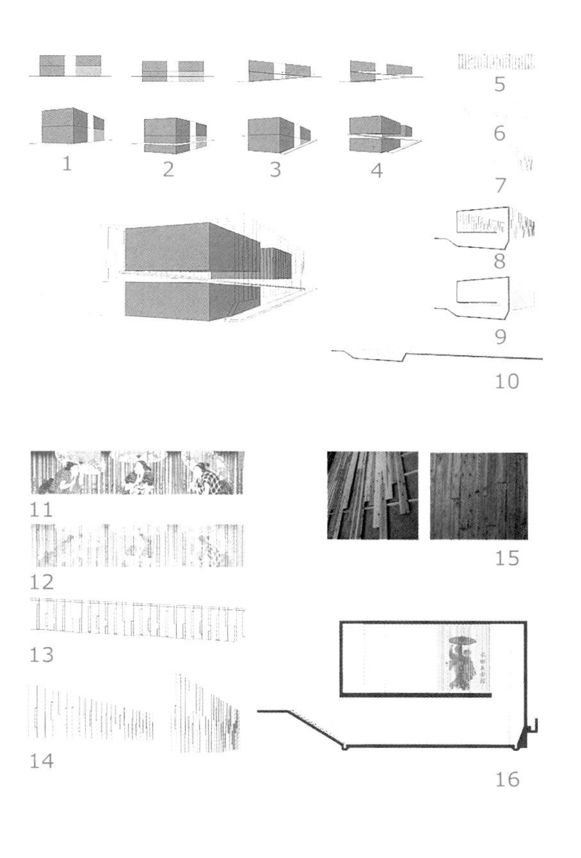

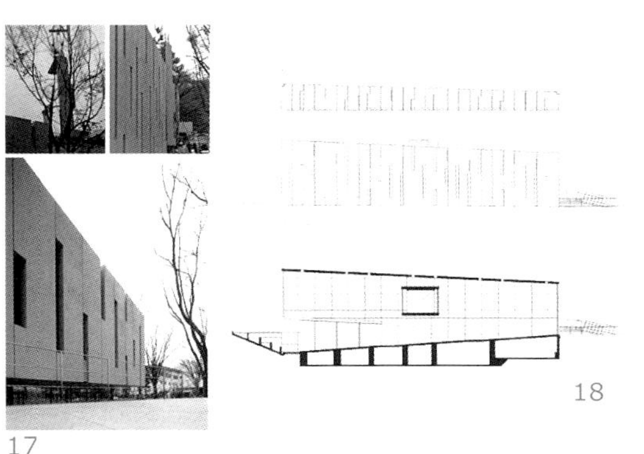

17

18

1. Stack 堆叠
2. Shift 转换
3. Ramp 斜坡
4. Float 漂浮
5. Pattern 形式
6. Tilt 倾斜
7. Screen 屏风
8. Slots 窄孔
9. Shell 骨架
10. Excavation 发掘
11. Ukiyo-e 浮世
12. Pattern 形式
13. Tilt 倾斜
14. Screen 屏风
15. Cast-in-place texture 浇入式质感
16. Section through gallery lounge 画廊休息室剖面
17. Installation of pre-cast pieces 安装预制模板
18. Elevations sections through ramp to information space 立面剖面

The building is like a hollow rectangle with facades in different lengths. When light is projected into the square windows of such a type, it will be vivid and that's vigor nature gives to architecture. Simple outline gives the building to show the natural beauty in a low-profile way. To harmonize with the surroundings, the façade is decorated with glass on one side and the landscape outside by accident becomes a flowing painting on the square glass.

该建筑如同一个镂空的长方体，立面上设计以长短不一、错落排布的条状长方形为主，当阳光射入这些条状的方形开窗时，光线会让空间产生灵性，这正是自然赋予建筑的生命力所在。简洁的造型设计让建筑以一种特别低调的方式来展现自然的美。为了与周围的环境协调，主建筑的一面以玻璃装饰，外面的景色不经意间成为了方形玻璃墙上一幅流动的装饰画。

For a simple rectangular shape, there is no extra space. The building in open design is not only well ventilated but looks quite imaginary. The randomly placed lines combine with the squares to create a façade of artistic beauty.

作为一个简洁的矩形形状，这个当代、标志性的房子没有多于累赘的空间。开放式设计，不仅通风，也让这个看起来规规矩矩的方块空间充满了想像。抽象的随机的线条与方形块的结合，营造出立面的艺术美感。

48

48

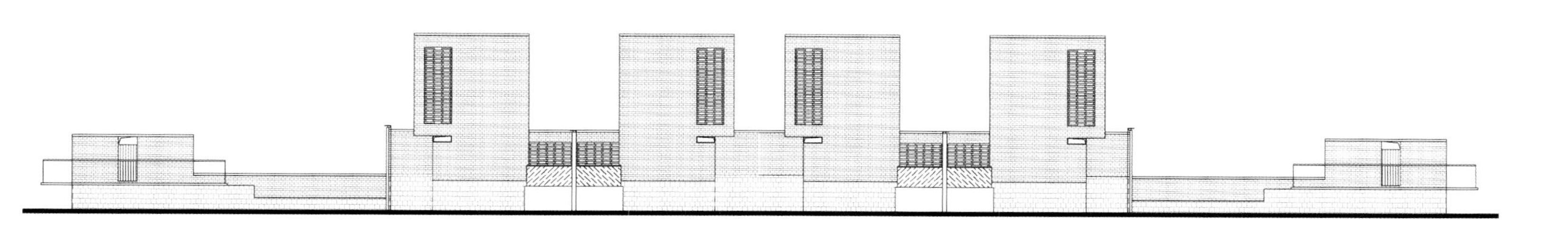

The Chinese complex in the square façade is that in the space form creation, the project, by means of multi-layer square courtyard overlapping, embodies the traditional front yard, central yard, back yard and side yard in the limited area, all of which form the controlled space rhythm.

方形表皮里的中国情结：在空间形态的营造上，项目通过多重方形庭院嵌套的格局，在有限的面积内将中国传统民居中的前院、中庭、后院和边院元素包含了进来，经过重新排布后，形成了收放有致的空间节奏。屋面的金属板与外立面的方形灰砖形成现代与传统的对话关系，这种关系通过悬挑、延伸、转折、起落，显得富有生机。在方形窗的安排上，建筑师并没有复制传统住宅中的木质花窗，而是通过石材切割与连接形成了漏窗的概念，体现出内敛的审美情趣。

THE VILLAGE
万科第五园

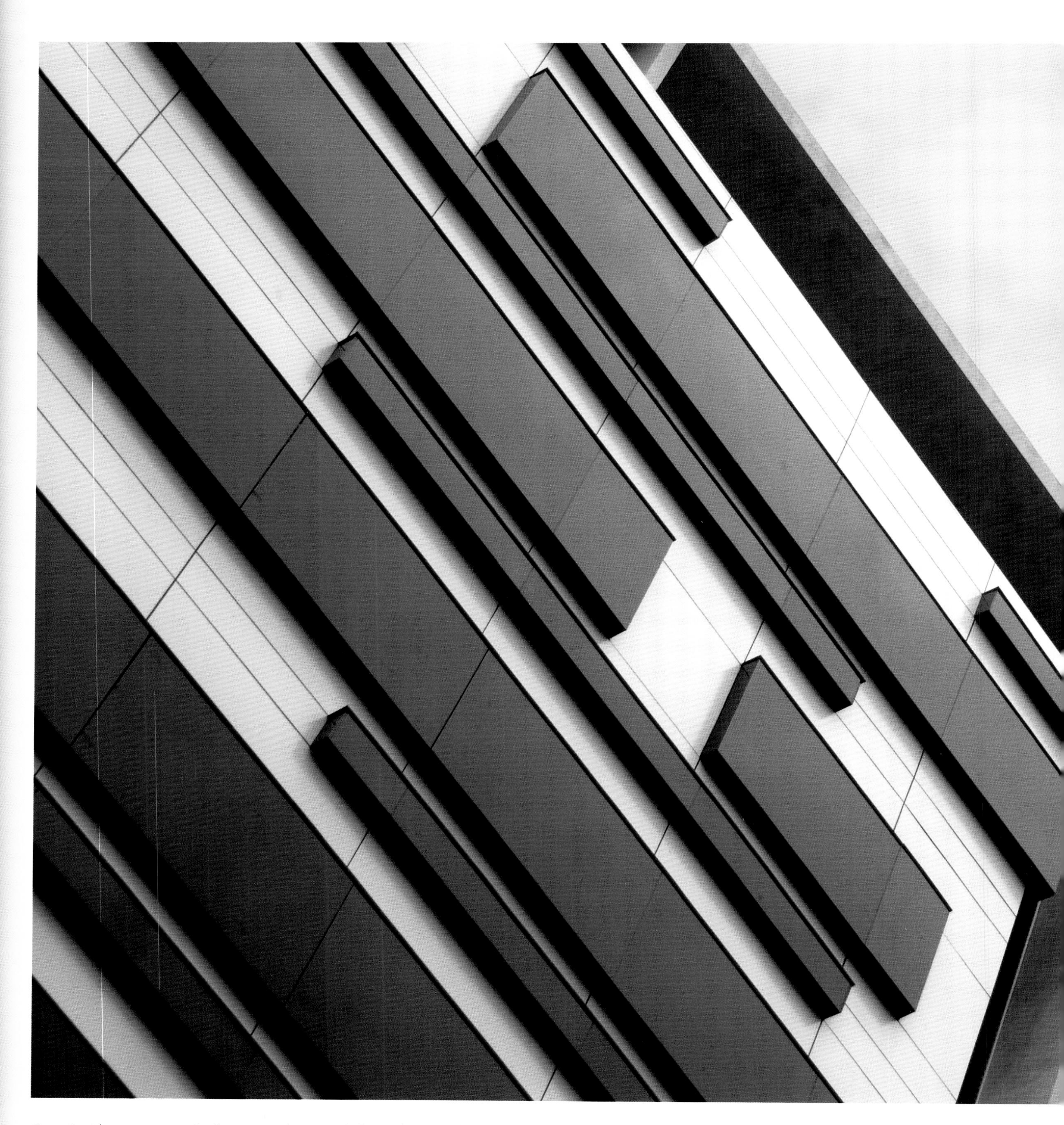

Due to the geometric form and material on the surface, the building is modern and unique in appearance, offering itself a shape different from others. The newly designed exterior wall unfolds continuously to set off the architectural surface in an artistic way.

因为表面的几何形状和材料，给了该建筑一个现代而独特的外观，使它区别于其他相类似的建筑物。外墙设计新颖，不同宽度的方形条带不连续地打开，把建筑表皮衬托得既简洁又富有艺术感。

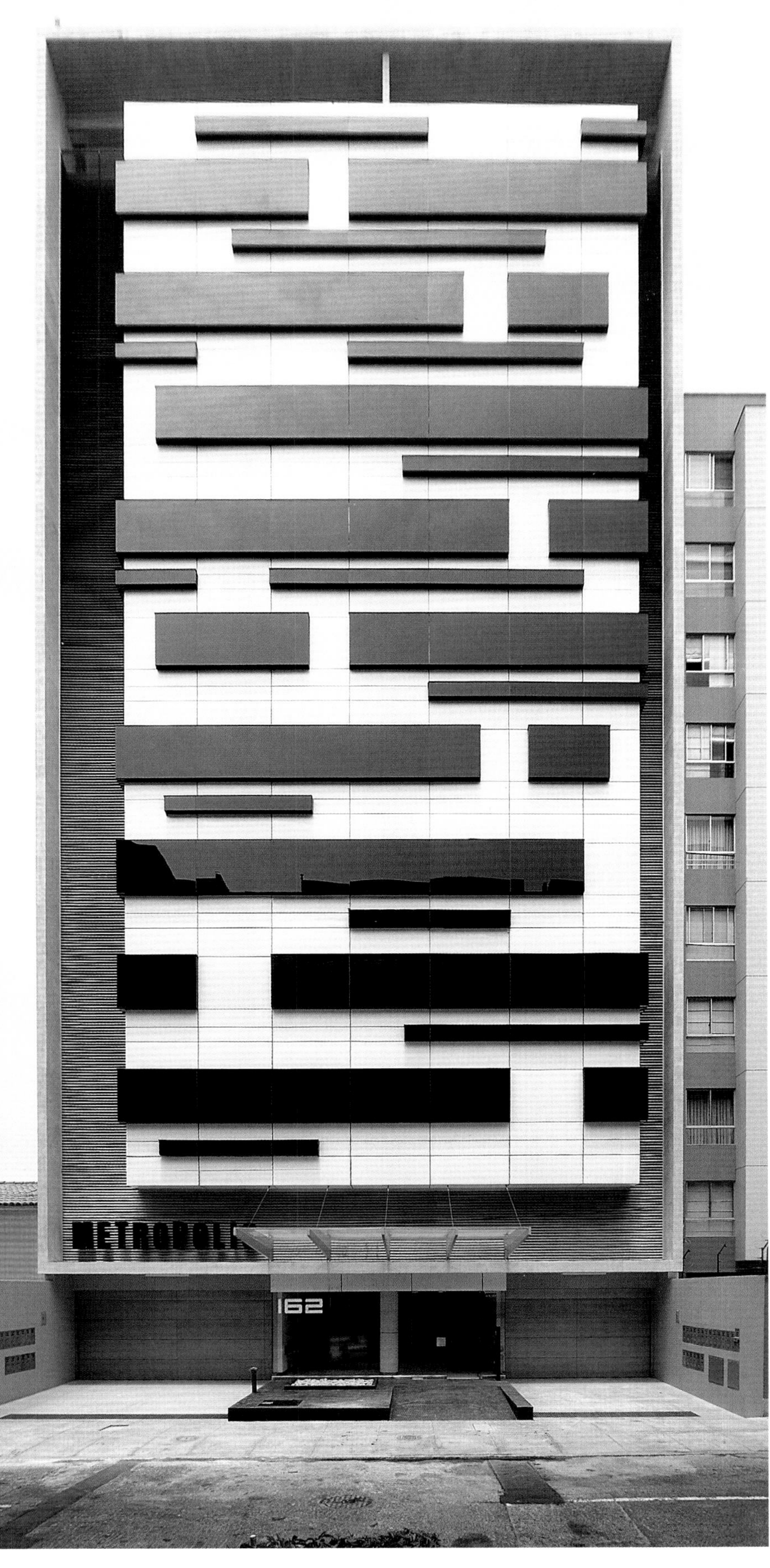
162

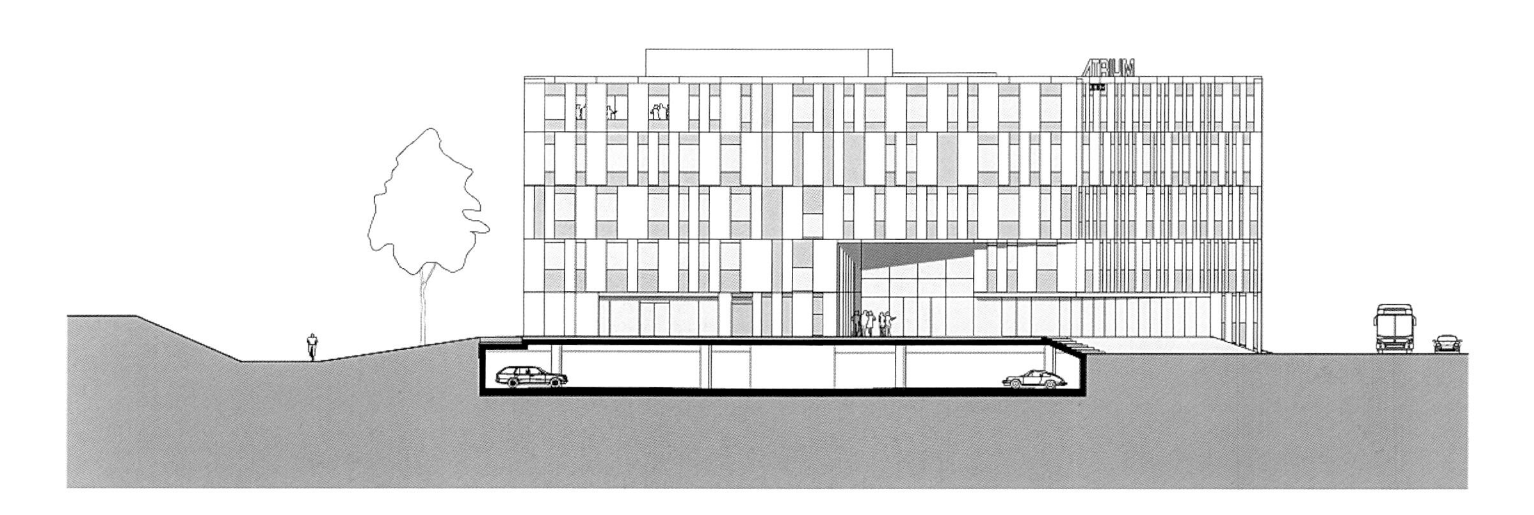

East View
东部视图

The façade is made up of pre-built concrete components interwoven into an irregular square lattice. Viewed as a whole, the black and white strewing breaks the cross section of the building to bring privacy and co-existence.

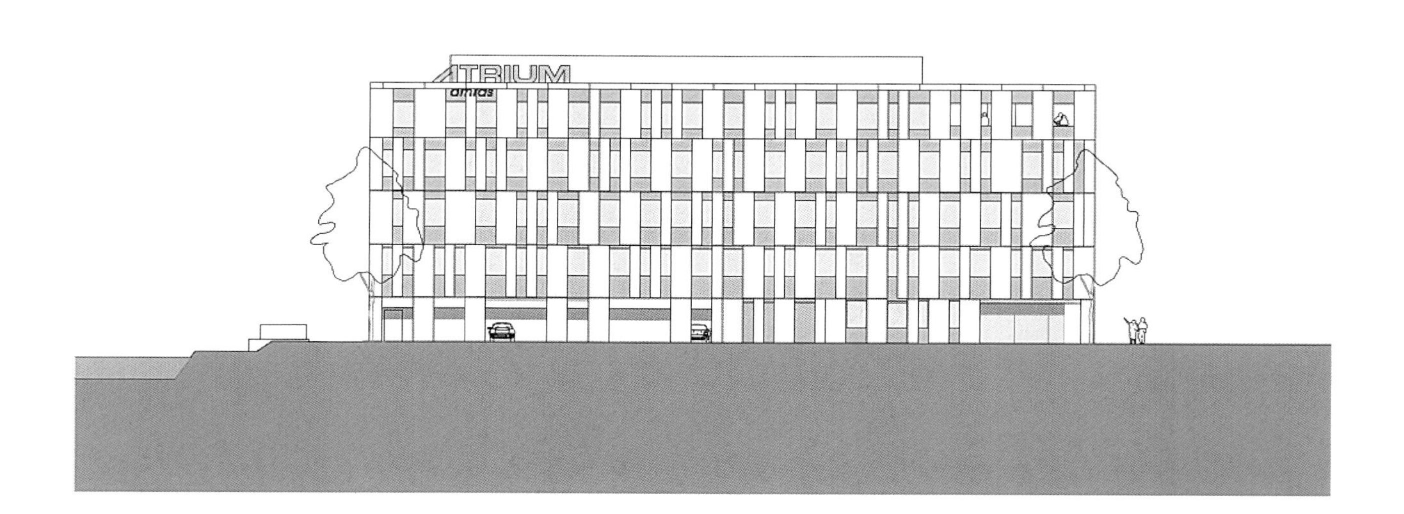

South View
南部视图

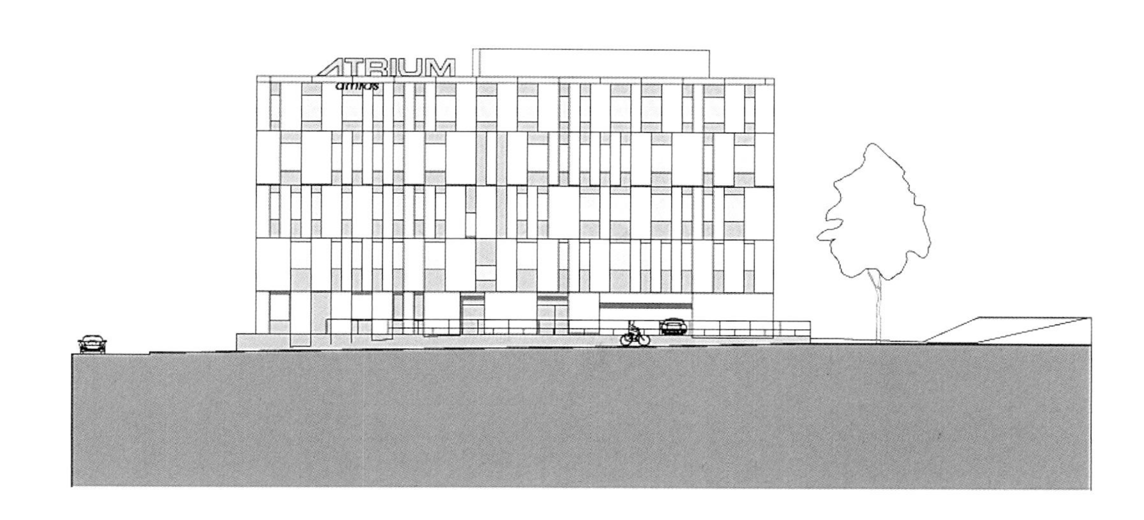

West View
西部视图

此建筑表皮由预制混凝土构件，交替在一个不规则的方形网格相间的窗洞口组合而成。从整体观感上看，黑与白的错落打开了建筑的横向立面，私密与开放共存，表现出十足的现代感。

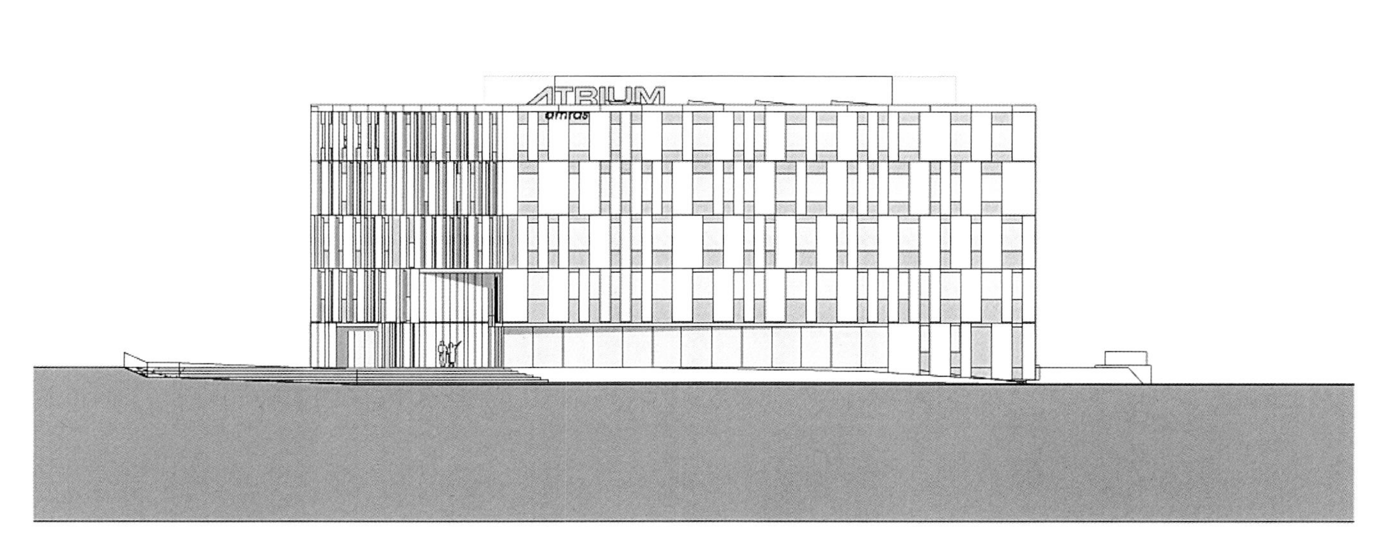

North View
北部视图

The spiral mosaic building: It adopts a façade that integrates wood with glass. The complex green building is in a spiral form. Although it has only three floors, every floor has access to a big air garden as if having a garden back yard. In particular on top, the design takes full advantage of the balcony to cover the roof with plants, regulating the weather conditions to have the effect of being cool in the summer and warm in the winter.

马赛克包围的螺旋建筑：外部立面是由天然木材板交替木结构玻璃面板上的水平网格。设计的这栋绿色植物层叠的综合楼，呈螺旋状。尽管只有三层，每层的设计都附上一大片的空中花园，仿佛每一层都像底层一样有花园式的后院。特别之处位于顶层，设计充分利用天台的优势，以植物全面覆盖屋顶，一来扩大了植物覆盖面，二来以天然的植物季节性调节，产生冬暖夏凉的温室调节效果。

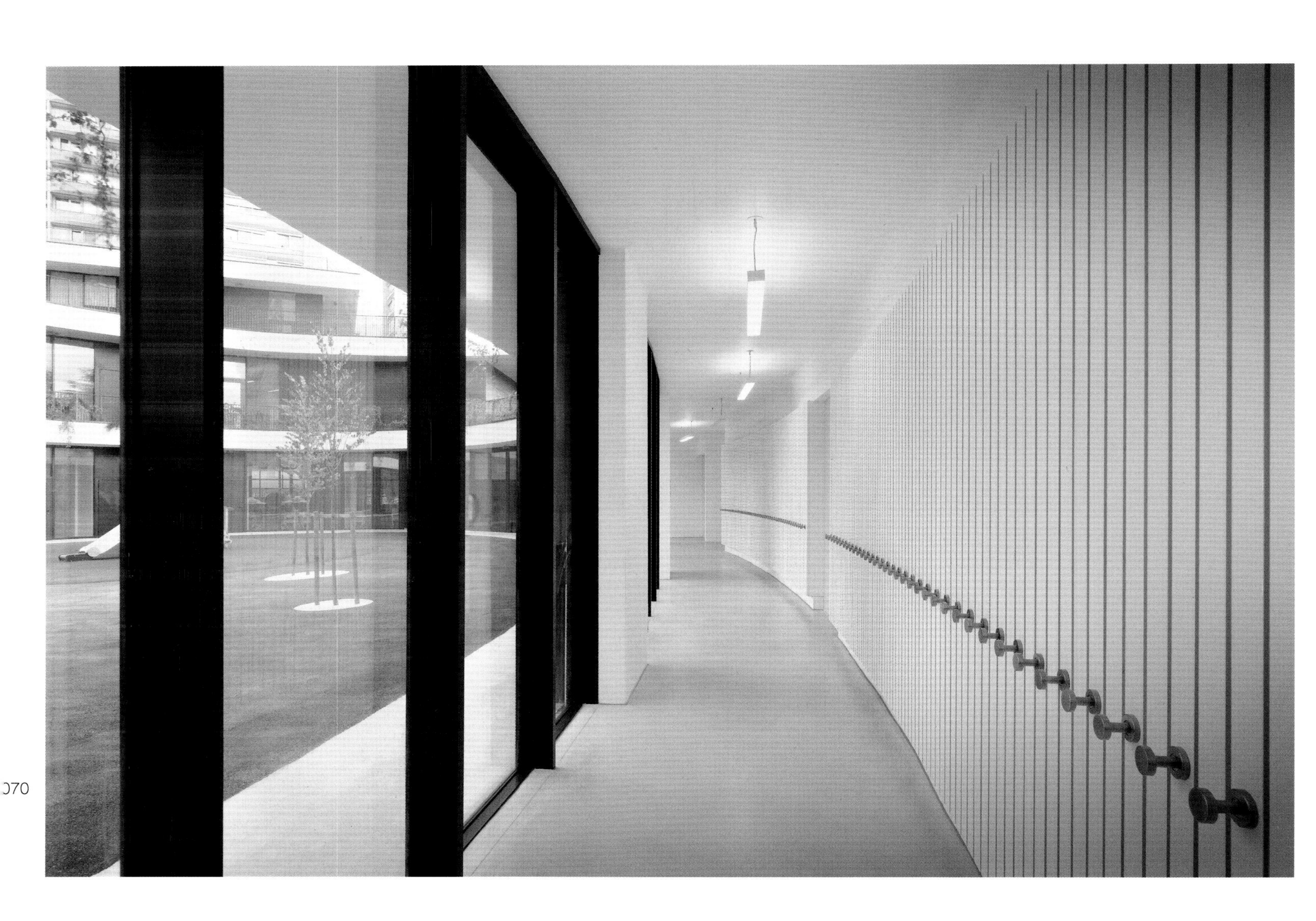

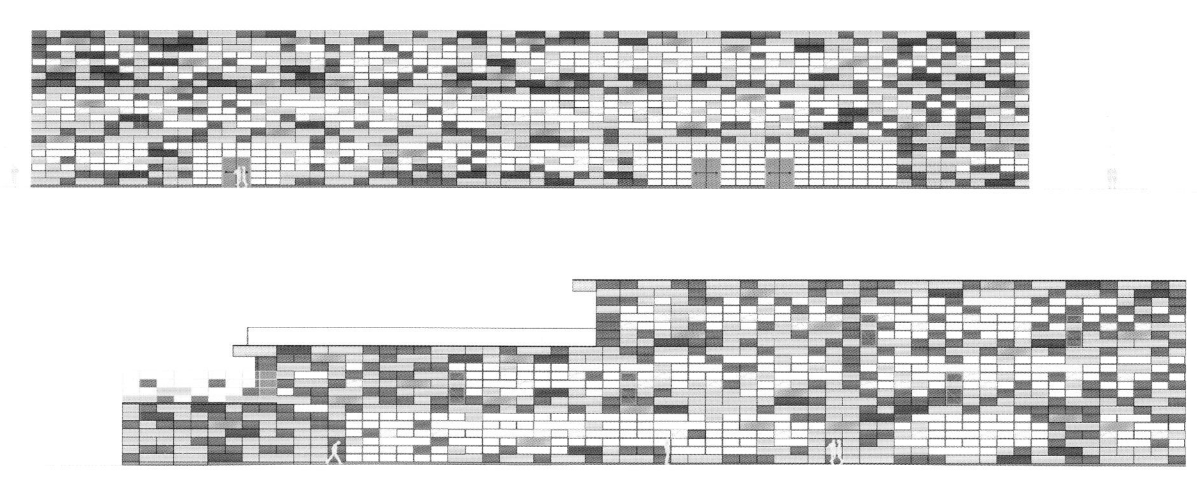

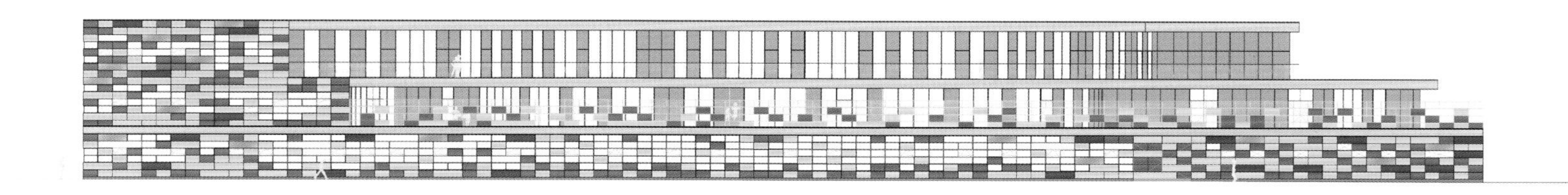

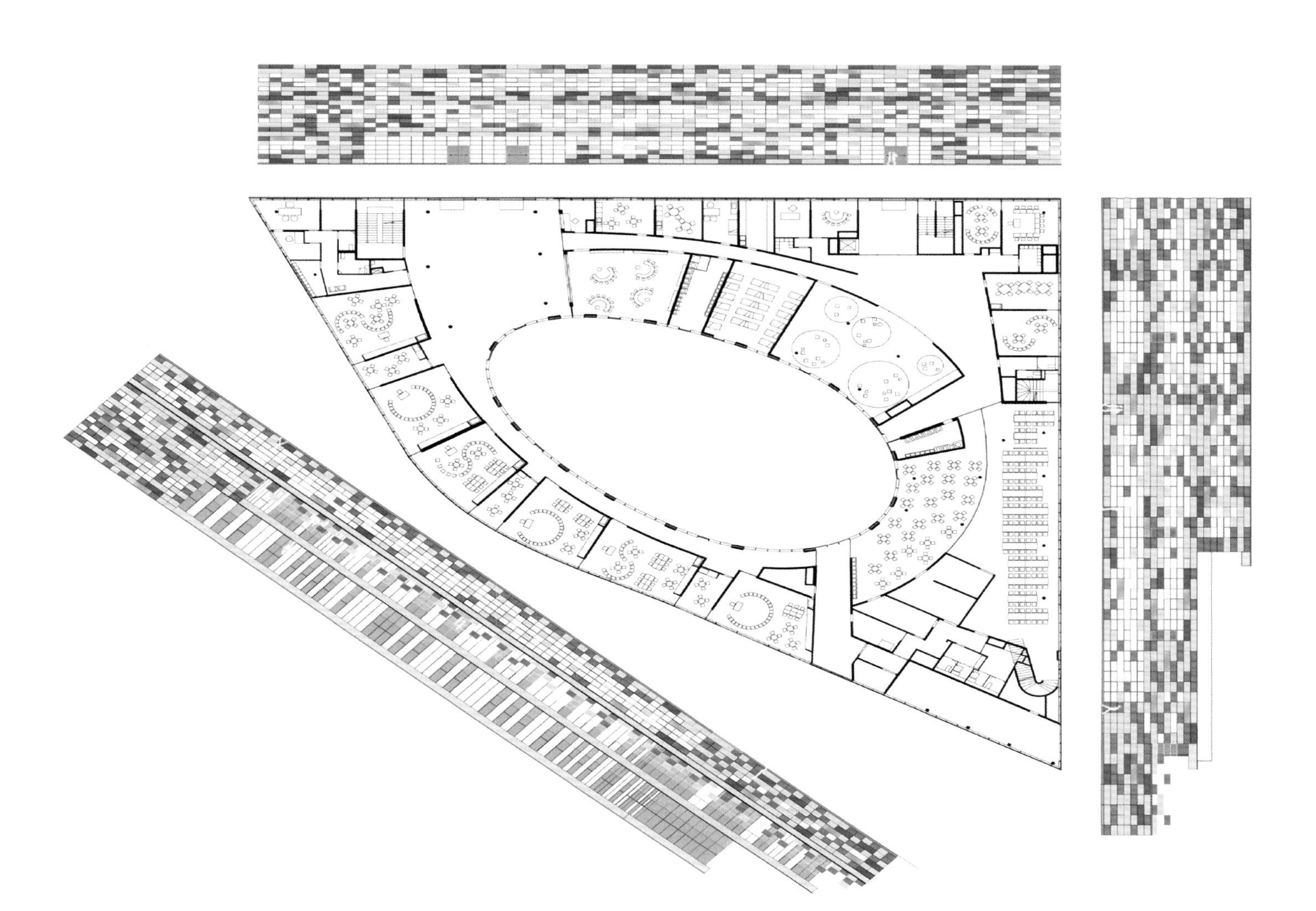

How does the uneven square combination display the abstract geometric spirit? Here, lighting decoration is the key and it deepens the contrast to make the square façade more creative and breaks the conventionality and tidiness of square planes. Different lighting effects are introduced to bring different atmosphere and experiences.

凹凸方形的组合又是如何表现抽象的几何精神？在这里，灯光的营造成了关键，它加深和增强了这种对比，使方形立面造型更加的别致而有创意，打破了方形平面一惯的规矩和整齐。利用不同光线的营造，增添了不同的氛围，带给人们不同的心理体验。

PURO
Hotel

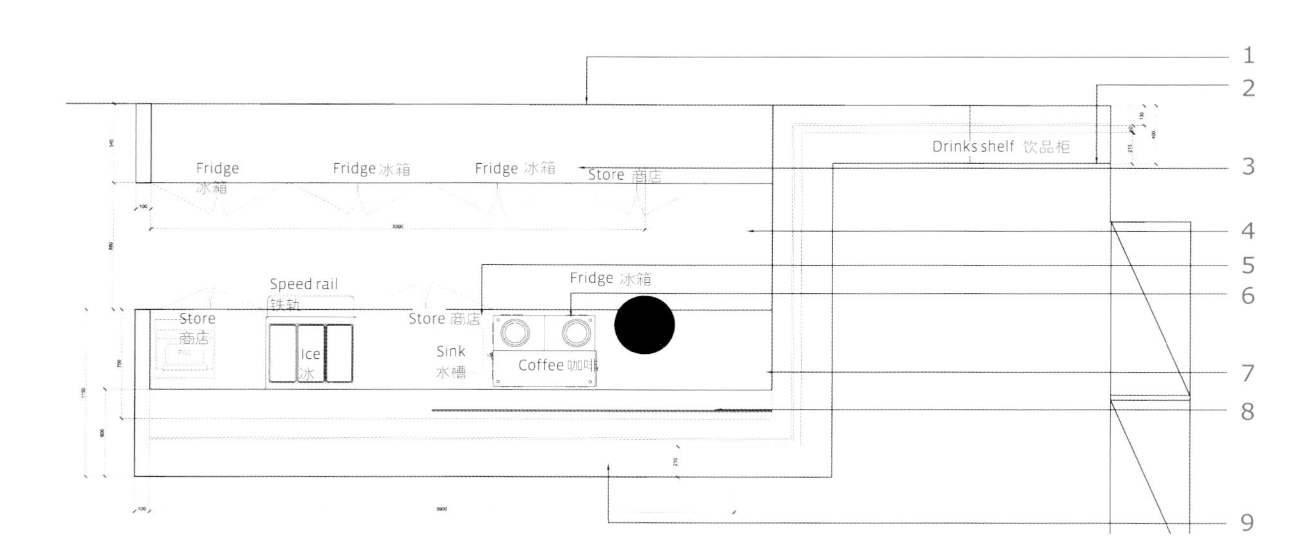

Bar Plan
酒吧平面图

1. Strip led to be recessed into mirror cladding behind back bar. See elevations for details 条状物凹进后吧台的涂层。详见立面细部图。
2. 100mm polished concrete work top & drinks shelf with 150mm return at cashier end of unit 100mm抛光混凝土顶板&饮品柜在收银台150mm位置
3. Strip led to be located to the underside of front bar to act as work light. 条状物放入前吧台底面以作为工作灯
4. Strip led to be located to the underside of the bar to illuminate bar front 条状物放入吧台底面以照亮前吧台
5. Front bar to have brushed stainless steel work top 前吧台照亮不锈钢顶板
6. Coffee machine and related equipment located at end of bar. Client to check equipment specification with joinery dimensions 咖啡机与相关设备放吧台末端。客户检查设备规格
7. Line indicates line of boxing below work top 盒式线条位于顶板下方
8. 300mm toughened glass up stand built into bar top 300mm钢化玻璃嵌入吧台顶部
9. 100mm polished concrete work top & drinks shelf with 150mm return at cashier end of unit 100mm抛光混凝土顶板&饮品柜在收银台150mm位置
10. 300mm glass up stand built into bar top. Ontractor to use concealed channel formed into concrete top 300mm玻璃嵌入吧台顶部。承包商使用隐藏式水管嵌入水泥顶部
11. 100mm polished concrete work top & drinks shelf with 150mm return at casgier end of unit 100mm抛光混凝土顶板&饮品柜在收银台150mm位置
12. Strip led to be located to the underside of the bar to illuminate bar front 条状物放入吧台底面以照亮前吧台
13. Internal carcass of unit to be fabricated from marine plyor stainless steel bar system. Front of unit to be clad in timber to match wall finish 内部框架由不锈钢吧台体系支撑。前部由木材覆盖同墙壁表面相映
14. 150mm tile skirting to match floor finish 150mm贴砖与地板材料映衬
15. Data and power requirements for pos terminal to be confirmed by client POS机终端由客户确定其数据与能源供应
16. All bar equipment to have brushed stainless steel finish 吧台设备都涂上不锈钢材质

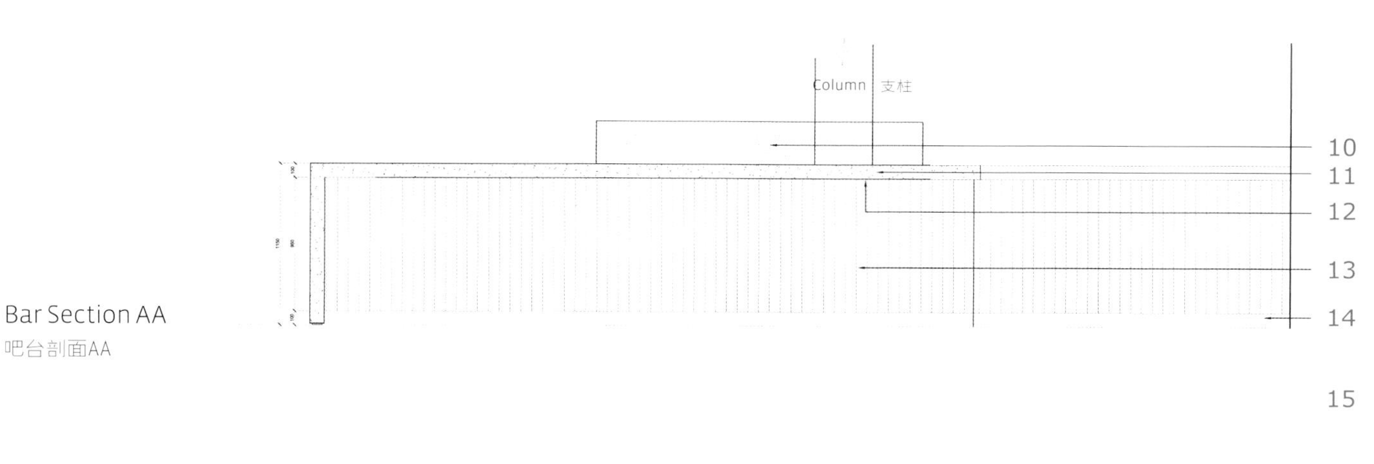

Bar Section AA
吧台剖面AA

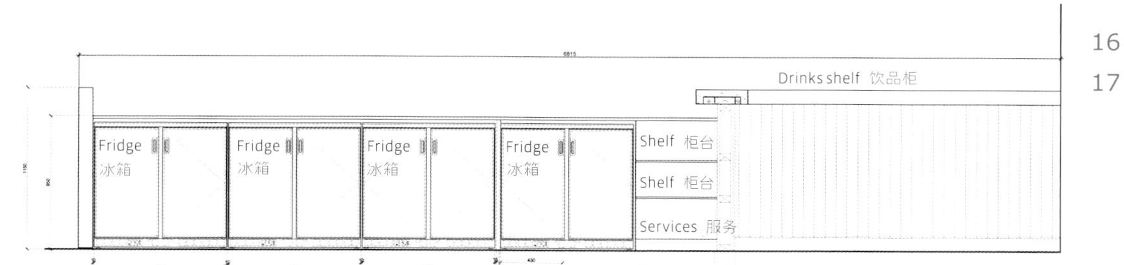

Bar Section BB
吧台剖面BB

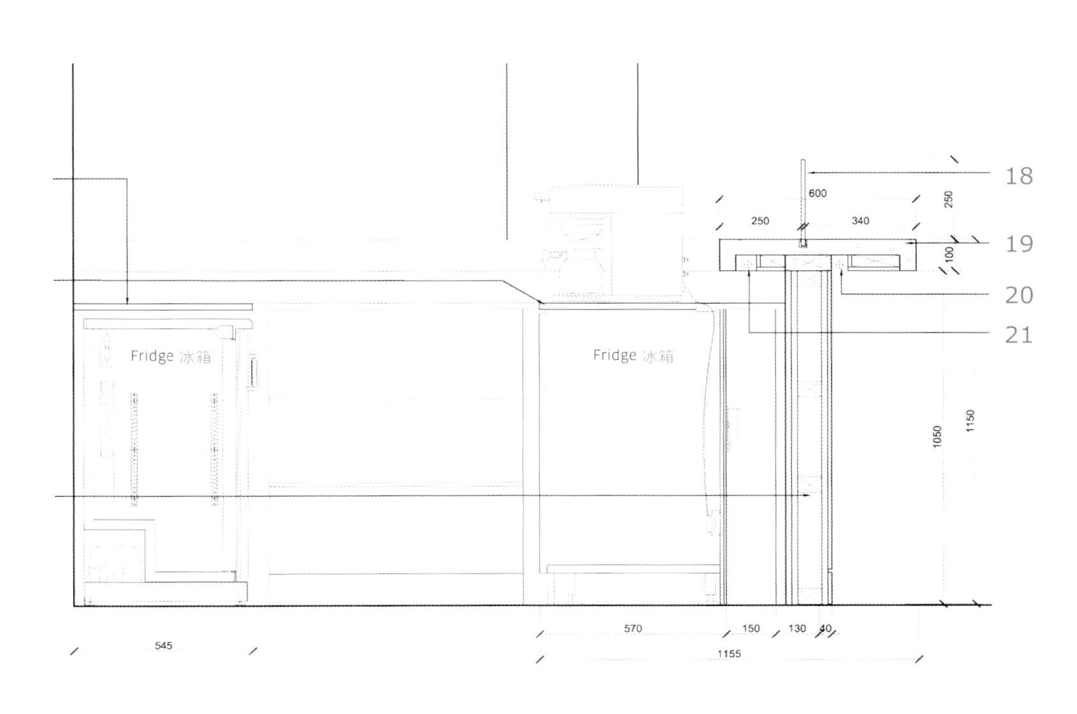

Bar Section CC
吧台剖面CC

17. All bar equipment to be specified prior to fabrication of bar unit. Contractor to ensure all equipment fits within prior to fabrication. 吧台设备按照吧台尺寸确定其规格。承包商确保设备符合要求
18. 250mm toughened glass up stand built into bar top 250mm钢化玻璃嵌入吧台顶部
19. 100mm polished concrete work top & drinks shelf with 150mm return at cashier end of unit 100mm抛光混凝土顶板&饮品柜在收银台150mm位置
20. Strip led to be located to the underside of the bar to illuminate bar front 条状物放入吧台底面以照亮前吧台
21. Strip led to be located to the underside of front bar to act as work light. 条状物放入前吧台底面以作为工作灯

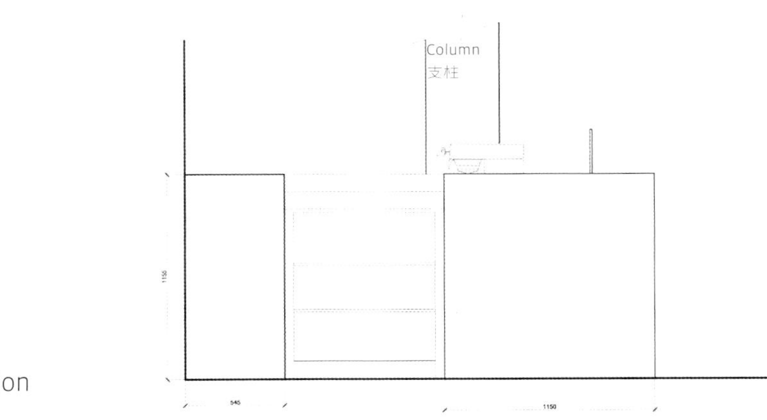

Bar End Elevation
吧台末端剖面图

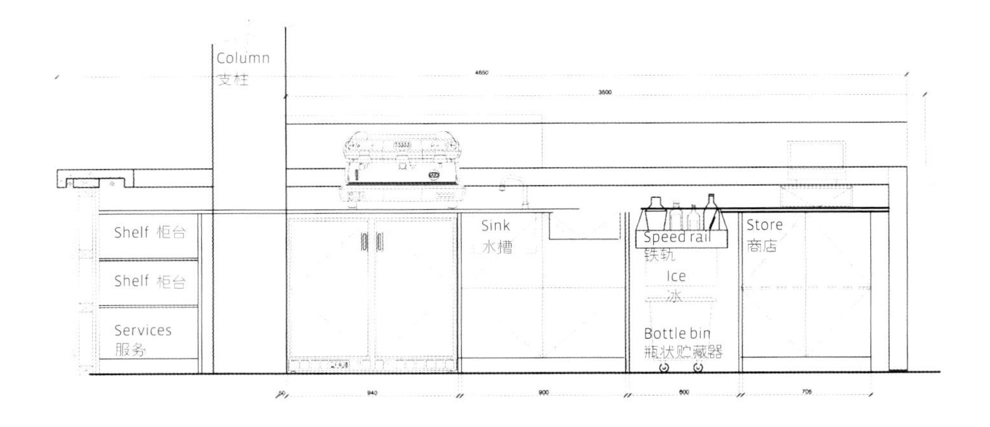

Bar Section DD
吧台剖面DD

The massive use of square planes makes the whole façade very modern through the contrast of white and black. The white wall, after simply being processed, makes the building smooth, natural, getting rid of dullness to seek the rich yet vivid architectural character.

建筑大面积运用方框的平面图形，利用黑白两色形成强烈的对比，使整体立面现代感很强。这里白色墙面突起的长方形，经过简单的处理和变形，使建筑简洁流畅、自然大方，破除单调感，从而求得丰富变化的造型效果，体现了规矩中又显灵动的建筑性格。

The repeated arrangement of squares on the glass façade can create a simple, penetrating plane environment and meanwhile reduces direct strong lighting and retains splendor.

THE
PUHUI
The charm is blooming

透明的玻璃立面上的长方形大面积重复排列，它能够营造出一种简约又通透的平面环境，同时，利用玻璃材质的透光性，使室内空间，受益于自然光过滤的方式，减少强光直射入内，同时又保持了壮观的景色。

THE
PUHUI
The charm is blooming

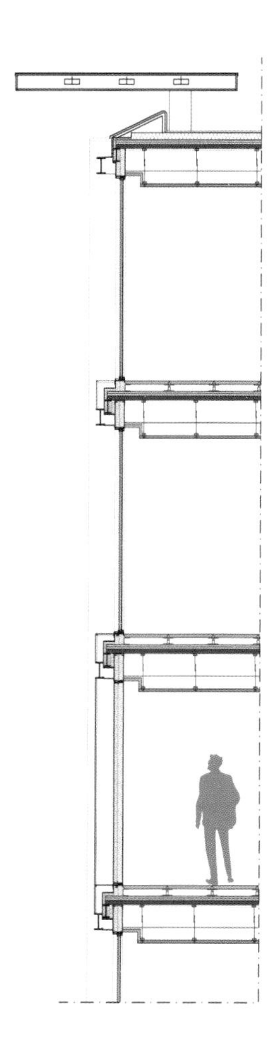

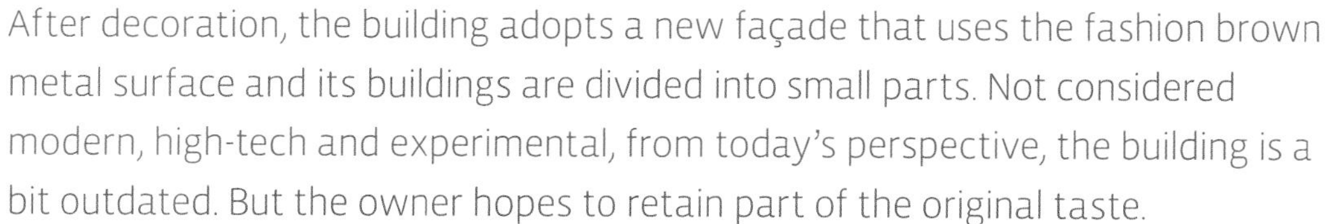

After decoration, the building adopts a new façade that uses the fashion brown metal surface and its buildings are divided into small parts. Not considered modern, high-tech and experimental, from today's perspective, the building is a bit outdated. But the owner hopes to retain part of the original taste.

Its façade becomes opener and more dynamic. The newly painted color makes the building shine and the piercing metal hole reflects a different atmosphere. Grey square glass is massively used and the surface reflects the surroundings and integrates the interior with the exterior, the past with the present.

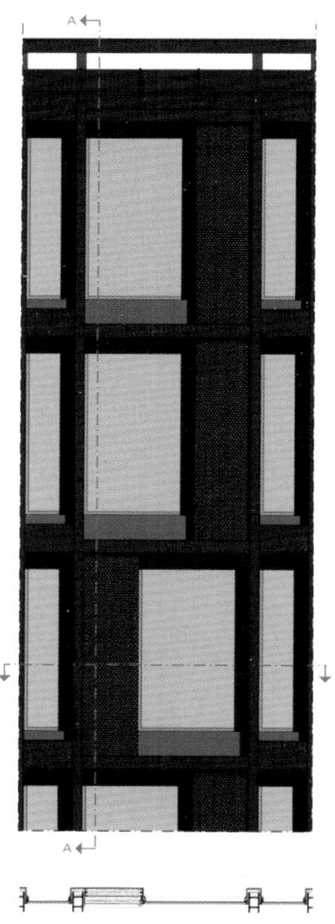

Prosp Frontale 正面图

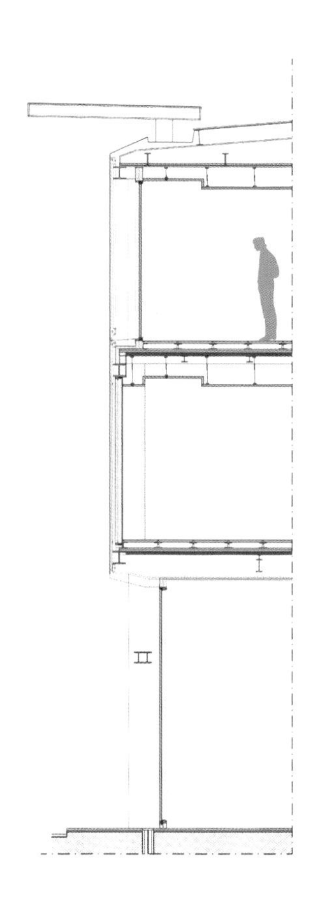

该建筑是经过重新整修的。新的立面采用了当时流行的棕色金属结构外皮，窗户被其分割成小块。与当时被认为的现代、高科技，还有试验性不同，从今天的眼光来看，建筑有些过时了。但业主希望建筑师在改建的时候，保留一点当时的味道。

外表皮与原来相比显得更开放和具备活力。重新刷上的颜色让建筑闪闪发亮，穿孔金属板在夜晚映射出另一种氛围和光芒。运用了大量的光滑灰色方形玻璃，它们的表面映射出周围的环境，联系过往与现在。

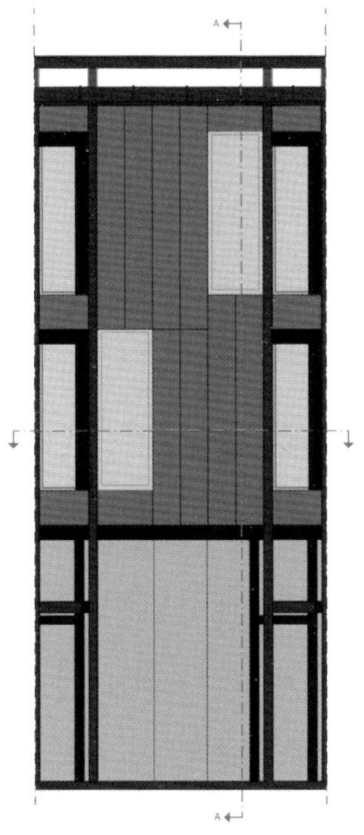

Prosp Laterale 侧面图

The project is to break through the old rules and to comply with the geometric shape of current blocks. The regular square plane of glass crosses from one block to another in an irregular way, taking on a dynamic beauty. The geometry planes forms a fixed alignment position and are supported by the steel framework. And the building takes on a triangle shape with two floors on its hanging structure, making the regular plane seem to be more flexible.

该项目设计的目的是从重建后的旧有规则中突破出来，服从现有街区的几何形状。全玻璃的规则的方形平面，以及其不规则的叠放方式，从一个街头横跨到另一个街头，它呈现出一种动态的几何美感。这些几何平面保持一个恒定的对齐方式，每个独立的实体都是由对角线或钢十字架支持着，这使得它自身非常稳定。最后，较高的建筑呈现出三角形的几何构图，悬臂上面共设置有两个楼层，它使得规整的方形平面更显灵活。

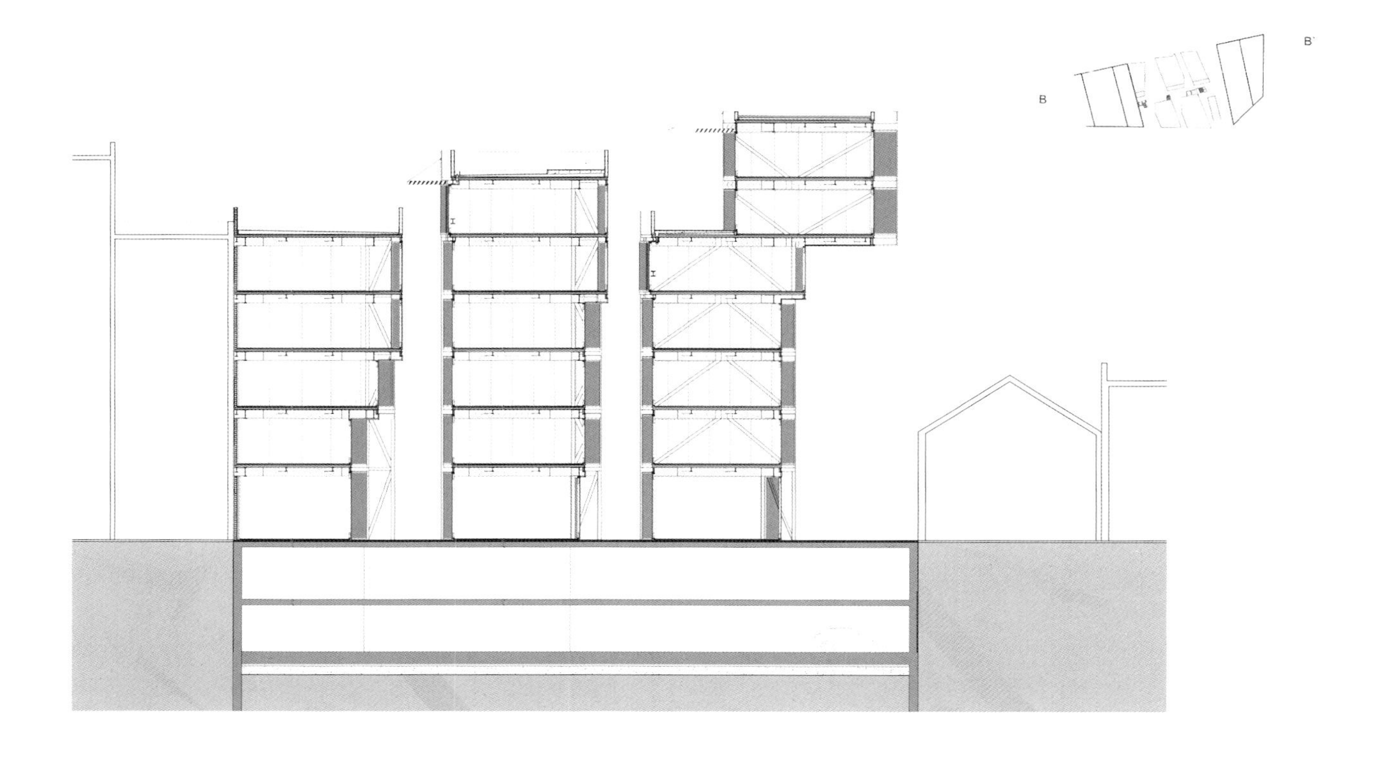
B
B'

COMARCAS CATALANAS, S. A.

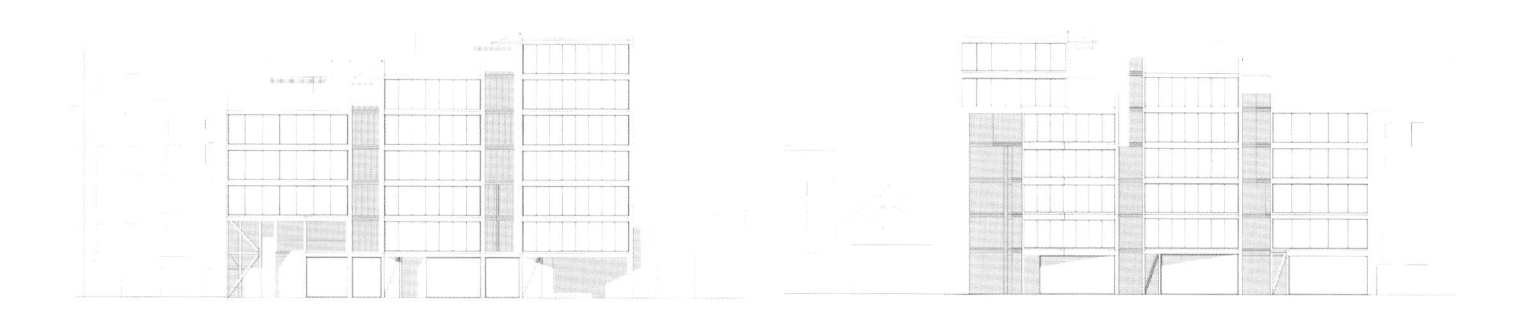

SMOOTH, FLUID, VIVID RHYTHM

流畅持续，富有节奏、律动的韵律感

For architecture, proper lines used can make the building good to see. The façade lines are sometimes interconnected, parallel, bumpy and uneven, giving a powerful sense just like happily blending gentleness with beauty.

The psychological feeling of linear lines is that the direction of a line plays great role in visual feeling. The vertical lines give gravity and balance; the horizontal lines steady feeling; the slanting lines dynamic. Different lines leave different impressions, creating simple, fashionable, elegant or gentle façade.

Lines can be divided into straight, curved, retroflex, free curve, etc. The vertical line is stiff; the horizontal line still; the slanting line positive; the curved line elegant; retroflex line unsteady; heavy line pointed, etc. Based upon different property of lines, the designer will offer the building different spirits.

In addition, lines moving in different directions give different impressions. Horizontal flowing lines show a smooth direction and continual space; vertical lines, gravitational feeling, just like the positively upwards lines; slanting lines, straightforward feeling and also striking shock. The close arrangement of lines gives different visual impressions. The lines thick or thin at different intervals will have different textual effects. In general, the lines in whatever way contain the power to tranquilize people.

Although the linear combination varies, it mainly depends upon the general conception the architect. Based upon the function requirement, area, architecture technology and local weather conditions, the building adopts the linear form consistent with its inner character and harmonizes with the surroundings, creating a totally new architectural image and bringing vigor to the city.

对于建筑物设计，运用合适的线条能使建筑赏心悦目。建筑立面的线条有时纵横交错；有时平行相交；有时前后错落；有时高低不均；有时韵律起伏。直线给人以力量感和稳定感，曲线柔美雄浑，两者结合，犹如一段情境相融的故事，给人以美的享受。

线形给人的心理感受：一条线的方位或方向在视觉感受方面起着较大的作用。垂直线给人重力感和平衡感；水平线给人稳定感；斜线给人以动态感；曲线则给人张力感和运动感。线形因其本身不同的形态和心理特征，能传达出建筑想要营造的氛围，创造出或简约或时尚或典雅或柔美的立面形象。

线形具体可分为：直线、曲线、折线、弧线、自由曲线等，垂直的线刚直，具有升降感；水平的线静止、安定；斜线飞跃、积极；曲线优雅、动感；折线不安定；粗线稳重踏实，具有前进感；细线锐利、速度又显柔弱感。设计师在具体设计中往往会根据线条的特质，赋予建筑不同的精神面貌。

另外，线在构成中，由于运动方向不同也给人不同的印象。左右流动的水平线，表现出流畅的形势和自然持续的空间。上下垂直流动的线给人产生力学自由落体感，它和积极的上升线形成对照，可产生强烈的向下降落的印象。由左向右上升的斜线，给人产生一种明快飞跃的轻松的运动感。由左向右下落的斜线，使人产生瞬间的飞快速度及动势，产生强烈的刺激感。线的紧密排列也会给人带来不同的视觉印象：线如按照一定的规律等距离排列会形成色的空间并产生出灰面的感觉；不同距离间隔排列，或线的粗细变化，将会产生不同的肌理效果。形状不同的线的等距离排列，将会产生凹凸感。总的来说，无论线形以什么方式呈现，都具有能使人平静或振奋的精神力量。

对于建筑平面上的线形组合虽然千姿百态，但主要取决于建筑师的总体构思，根据所设计对象的功能要求与面积、建筑技术条件和当地自然气候与建筑环境等，采取与建筑功能和内在性格相一致的线形形式，并与周围的多种环境要素相协调，创造出新颖别致、与众不同的独特建筑形象，给建筑和城市带来生机。

Horizontal Line – The horizontal line is stretched, peaceful and quiet. Generally, the crossing lines on the façade contrast with the wall. Such lines are balanced and easy to adjust to the surroundings. Among them, horizontal lines are spacious, steady and wide, helping recall the ground in connection with peaceful, solid and quiet feelings. Simple lines can show the beauty of the façade. Repeatedly placed horizontal lines give steady feelings and simplified beauty, making you feel spacious.

横线条构图——横线条给人舒展、宁静、安定的感觉。一般立面上横向的水平线条与墙面形成强烈的明暗、虚实对比的效果。横线条具有某种平衡性，较易适应环境。其中，水平线有宽阔、平稳、延展空间之感，能使人联想到地平线与大地，引申为平实、牢固、安静等心理感受。简单的线条能够展示立面的美感，几根简单的横向线条会给人轻快、舒展开阔、简约、活泼之感，多条重复排列的横向线条给人平稳、安定之感。横向线条的简约美，可以使人觉得轻快明了，同时视野开阔。

D-CUBE DEPARTMENT STORE
Premiere
OPEN
CCTV

To honor industry history and modern environmental art pieces relations, the building is designed to be a modern complex integrating nature and cultural existence. The façade with marbles and crossing curves builds circuitous corridors take on the perfect blend with nature through permeability and elegant outlines of the building.

TNGT
ZARA

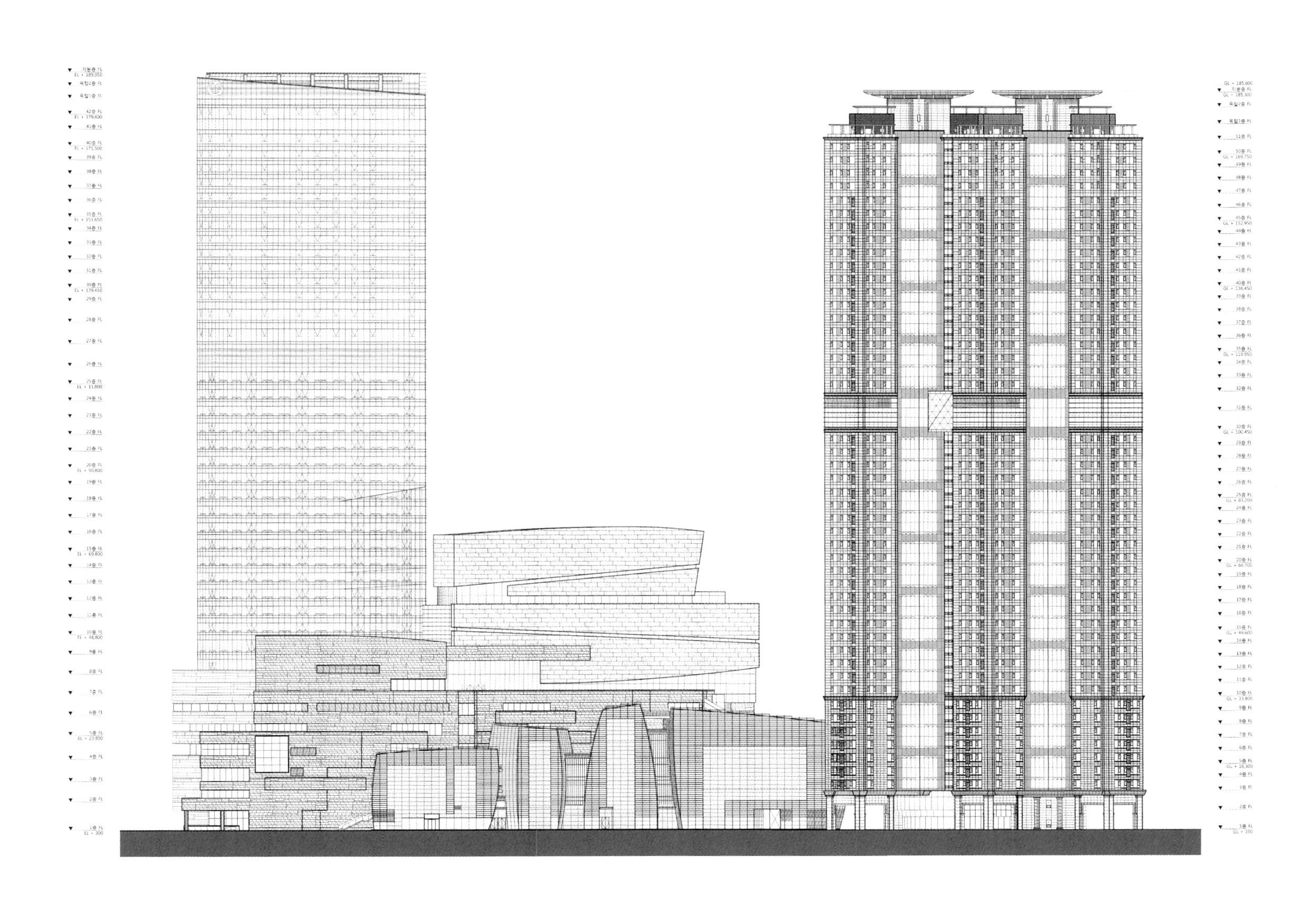

为了纪念工业历史和现代环境艺术作品之间的多元文化联系，该建筑被设计成为一个自然与文化共存的现代城市综合项目。外观以大理石为主要原料配合横竖交错的曲线轮廓的设计理念，打造了意大利小巷子感觉的迂回型走廊以及表现地中海天空的顶棚。立面的透光性以及优美的外框线条呈现了这座混合建筑与自然的完美衔接。

D-CUBE CITY
뮤지컬 맘마미아!
절찬리 공연중!
디큐브아트센터
SIEG

T.G.I. FRIDAY'S
ZARA

The vertical and horizontal crossings, big block and small block combination, and match of line and volume make any space interesting in small space. The bold lines connect the building in all parts and show the tension and firmness of the building to avoid dullness of the façade.

横竖线条的交错，大方块与小方块的组合，线条与体块的搭配，丰富的手法组合在一起使得任何一个小空间都变的丰富有趣。该建筑表面大胆地使用了大量的横竖向条纹，既在外型上将建筑的各部分联系起来，同时也表现出建筑的张力与稳重感，使立面避免单调呆板，具有较活泼的效果。

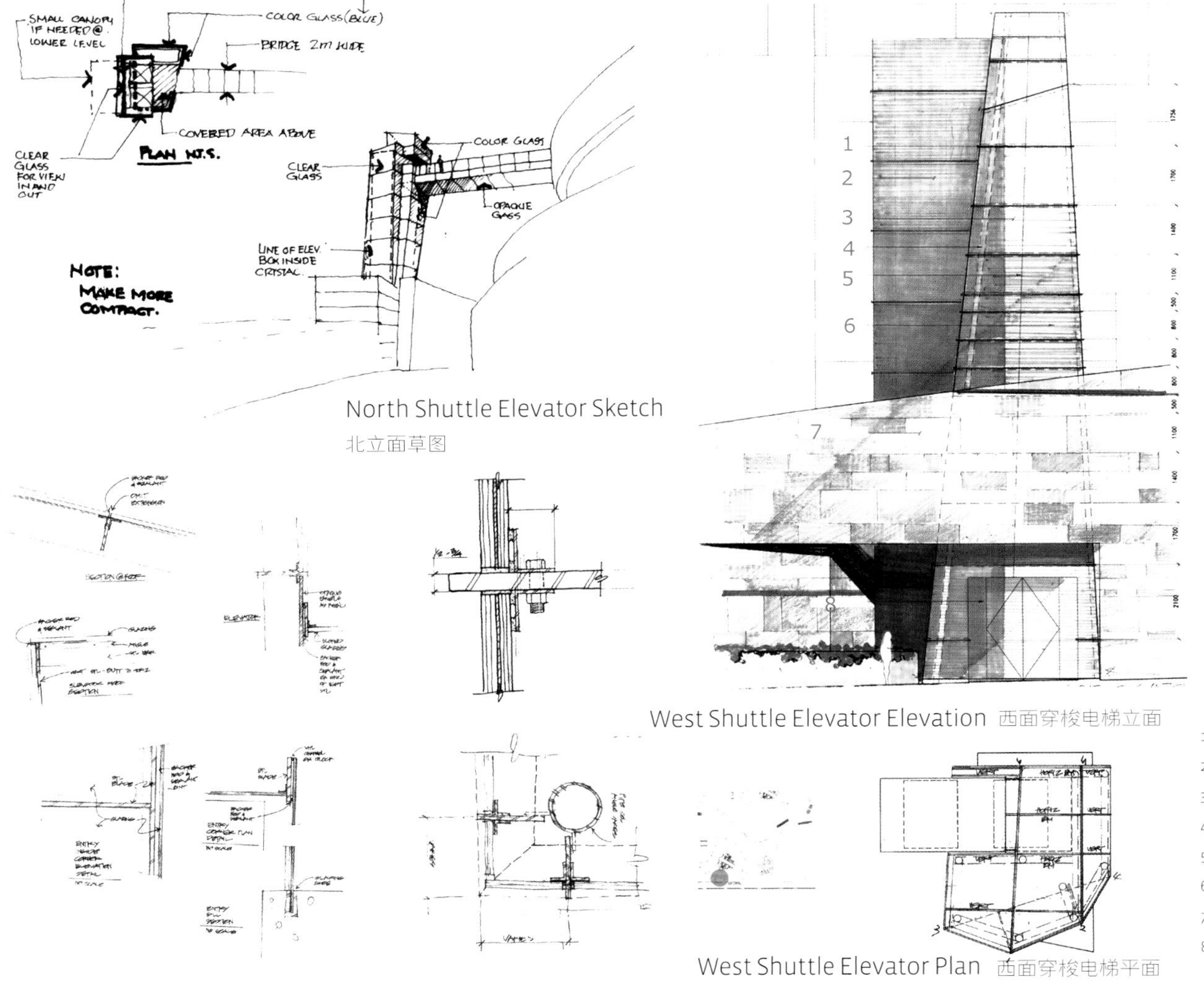

North Shuttle Elevator Sketch 北立面草图

West Shuttle Elevator Elevation 西面穿梭电梯立面

West Shuttle Elevator Plan 西面穿梭电梯平面

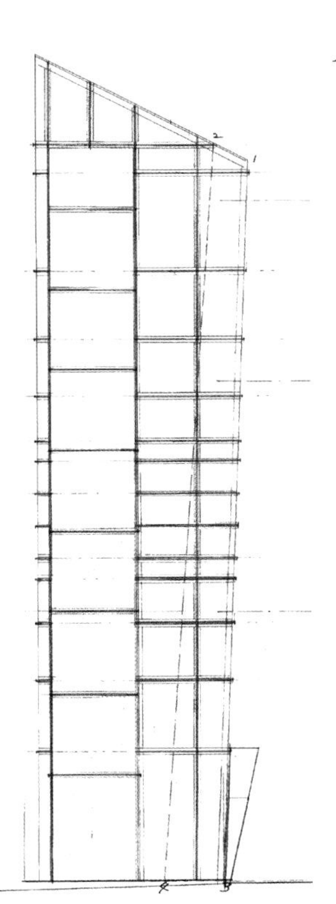

1. Painted Metal Frame Behind 涂料金属框
2. Clear Glass Elevator Beyond 透明玻璃自动扶手
3. Colored Glass 彩色玻璃
4. Painted Wetal Vertical Support Behind 垂直支持
5. Vertical Butt Joint 垂直对接
6. Edge Of Clear Class Elevator Beyond 观光电梯
7. Stone Wall In Front 前部石墙
8. Resident Entrance 住宅入口

The rhythmic trapezoid facade: The architect designs a piercing house of height distinction, with good ventilation and the house, the garden and the road form a unique community, helps recalling childhood memory. The northeast corner lifted and the southwest corner reduced permit sufficient light and air into the central courtyard.

富有曲线韵律感的梯形外表皮：建筑师设计了一个长而贯通、高度差明显的房子，采光好，与小花园和小路一起形成了一个独具特色的社区，不禁让人的思绪飘入南欧大山深处以及儿时家园的记忆中。东北角抬升，而西南角下降，使得光线与空气能够进入中央的庭院。

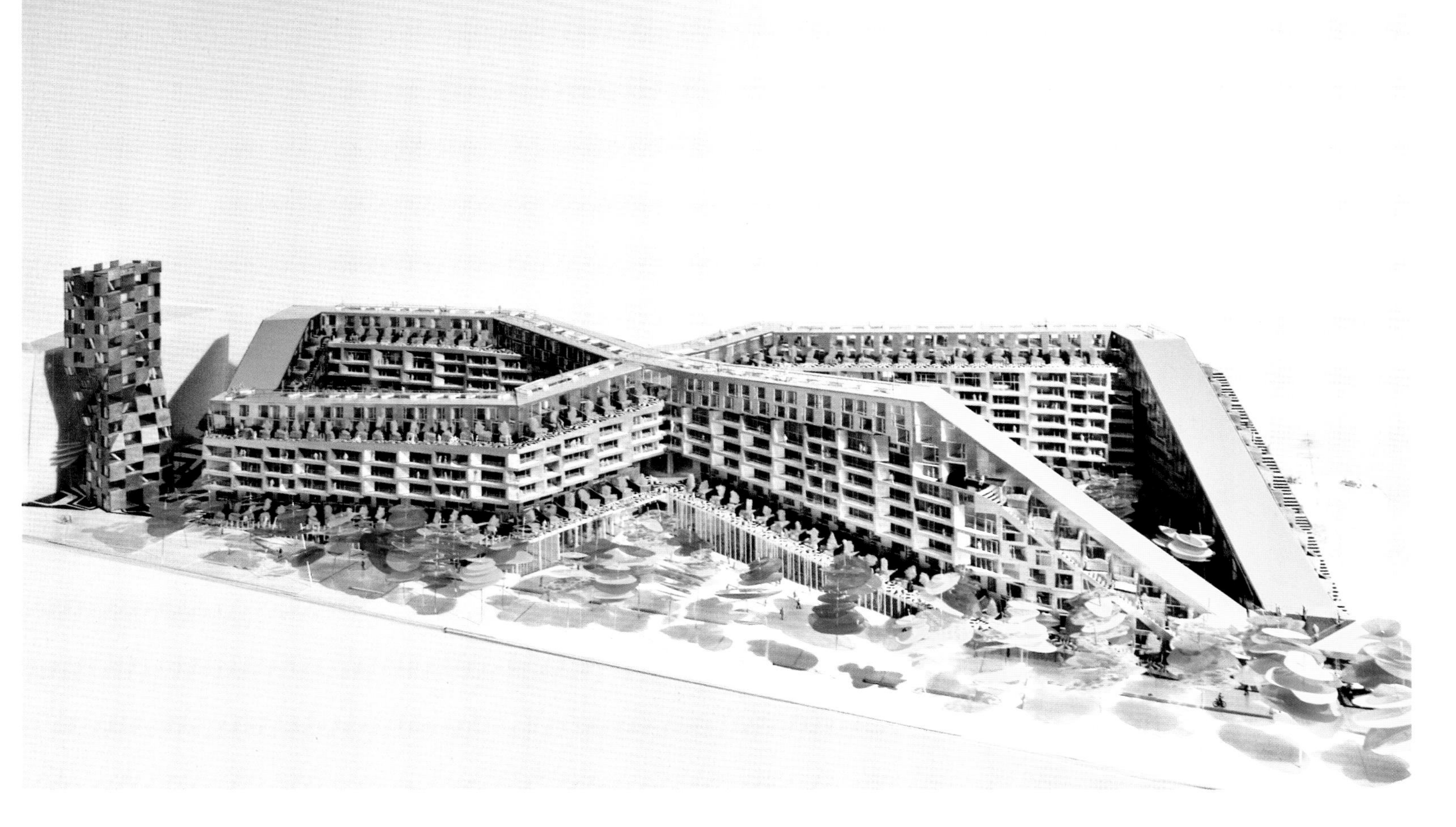

Vertical Line – The vertical line makes the whole building more eminent and arranged in a dotted way and improves its privacy and meanwhile gives the building intelligence. The vertical lines are often straightforward, succinct, steady and rigid. The vertical lines are simple, comfortable, sharp, etc. And the vertical lines in different lengths are orderly and give you a space of comfort and peace. The vertical lines are though succinct and lucid but they are not too dull. For modern cities, the design is suitable to modern people.

竖线条构图——竖向线条在视觉效果上会使整体建筑更加俊美挺拔，高低错落的组合排列，在提高私密性的同时更赋予建筑些许的灵动。竖线条常给人直截了当、干脆明快、坚实稳重、刚劲挺拔的观感。条状竖线给人的心理印象是简单、利落、舒适，有直接、锐利、明快、简洁等特性。大小均匀重复排列的竖向直线，给人以自然的延伸感；长短不一的竖向直线的排列，具有秩序感，在心理上让人感觉自由整齐又不失灵活，能够营造一种舒适、宁静的空间氛围。竖线虽然给人一种简练、简单明快的感觉，但不会让人感觉太单调，对于现代化城市，这种设计很符合当代人追求简单的理念。

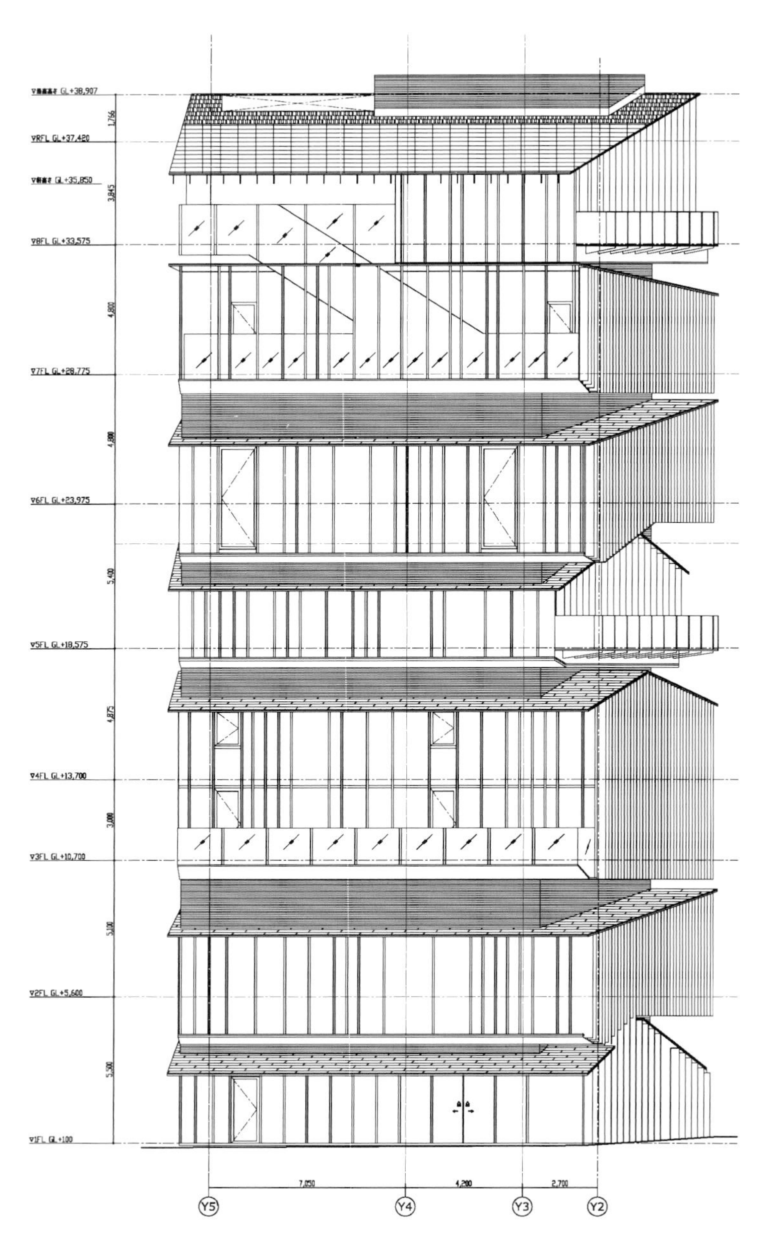

North Elevation 北立面图

Vertical Line – The vertical line makes the whole building more eminent and arranged in a dotted way and improves its privacy and meanwhile gives the building intelligence.

竖线条构图——竖向线条在视觉效果上会使整体建筑更加俊美挺拔，高低错落的组合排列，在提高私密性的同时更赋予建筑些许的灵动。竖线条常给人直截了当、干脆明快、坚实稳重、刚劲挺拔的观感。

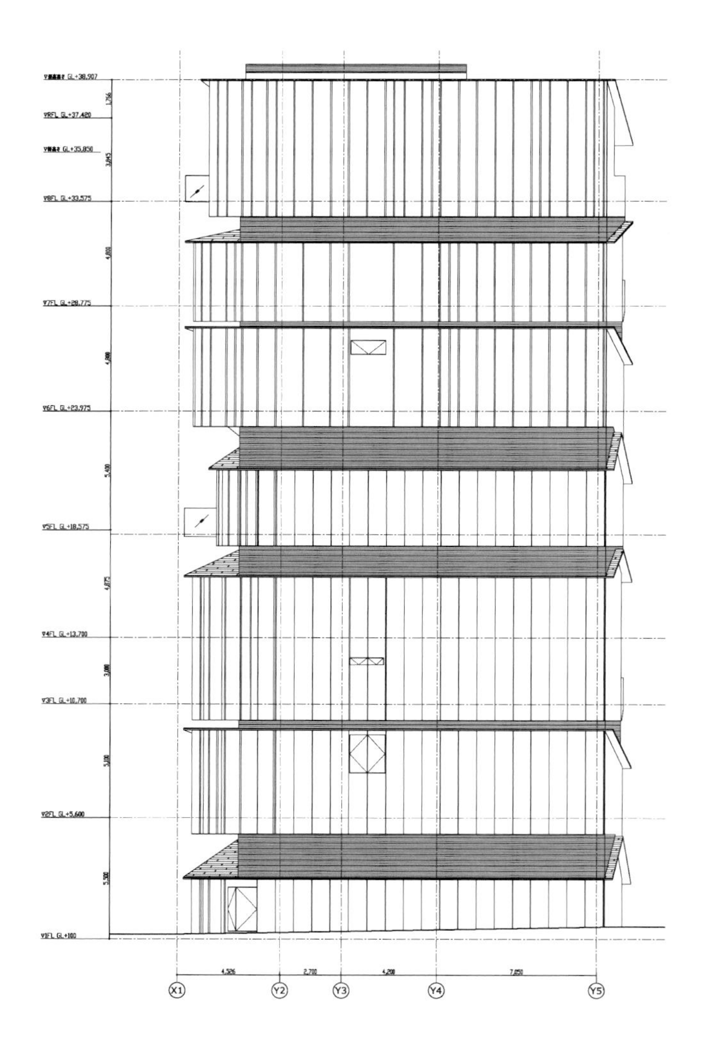

South Elevation 南立面图

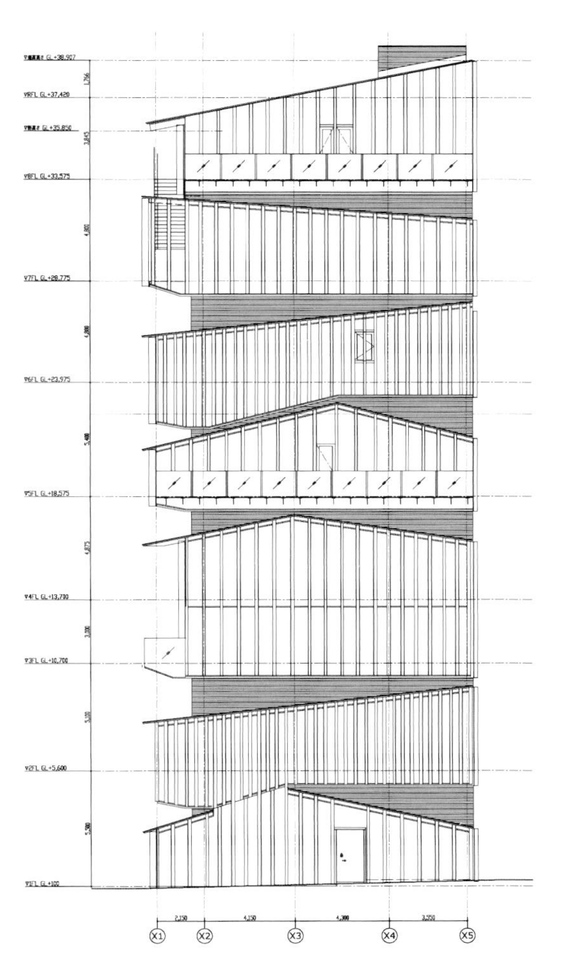

East Elevation 东立面图

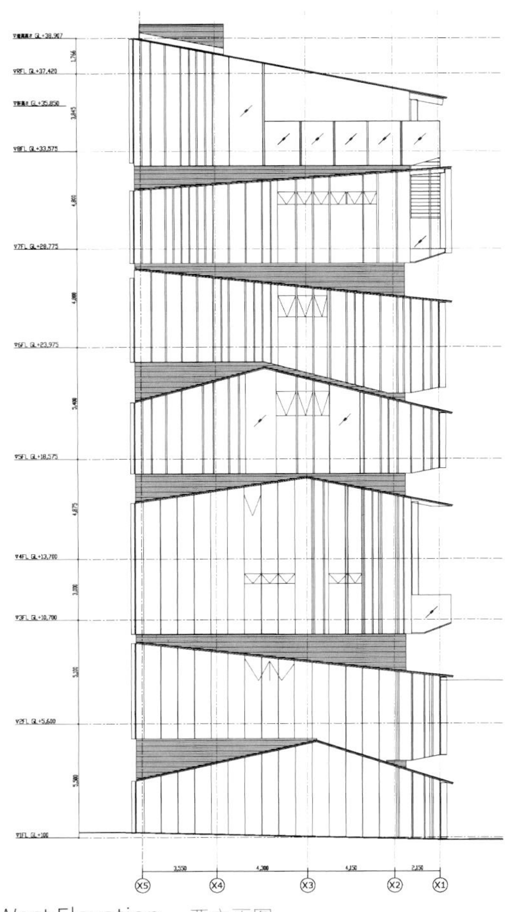

West Elevation 西立面图

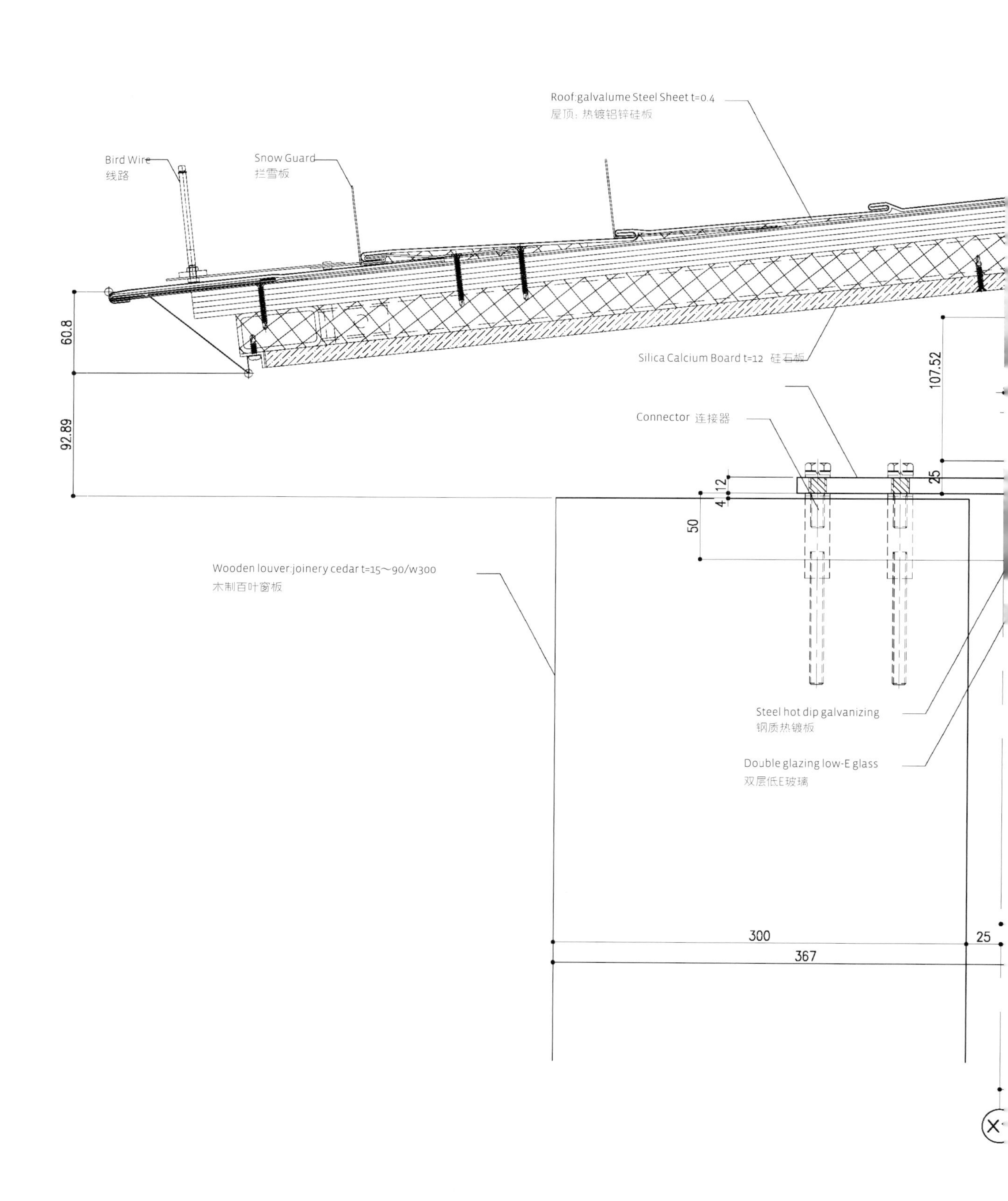

Wall Section 1 外墙剖面1

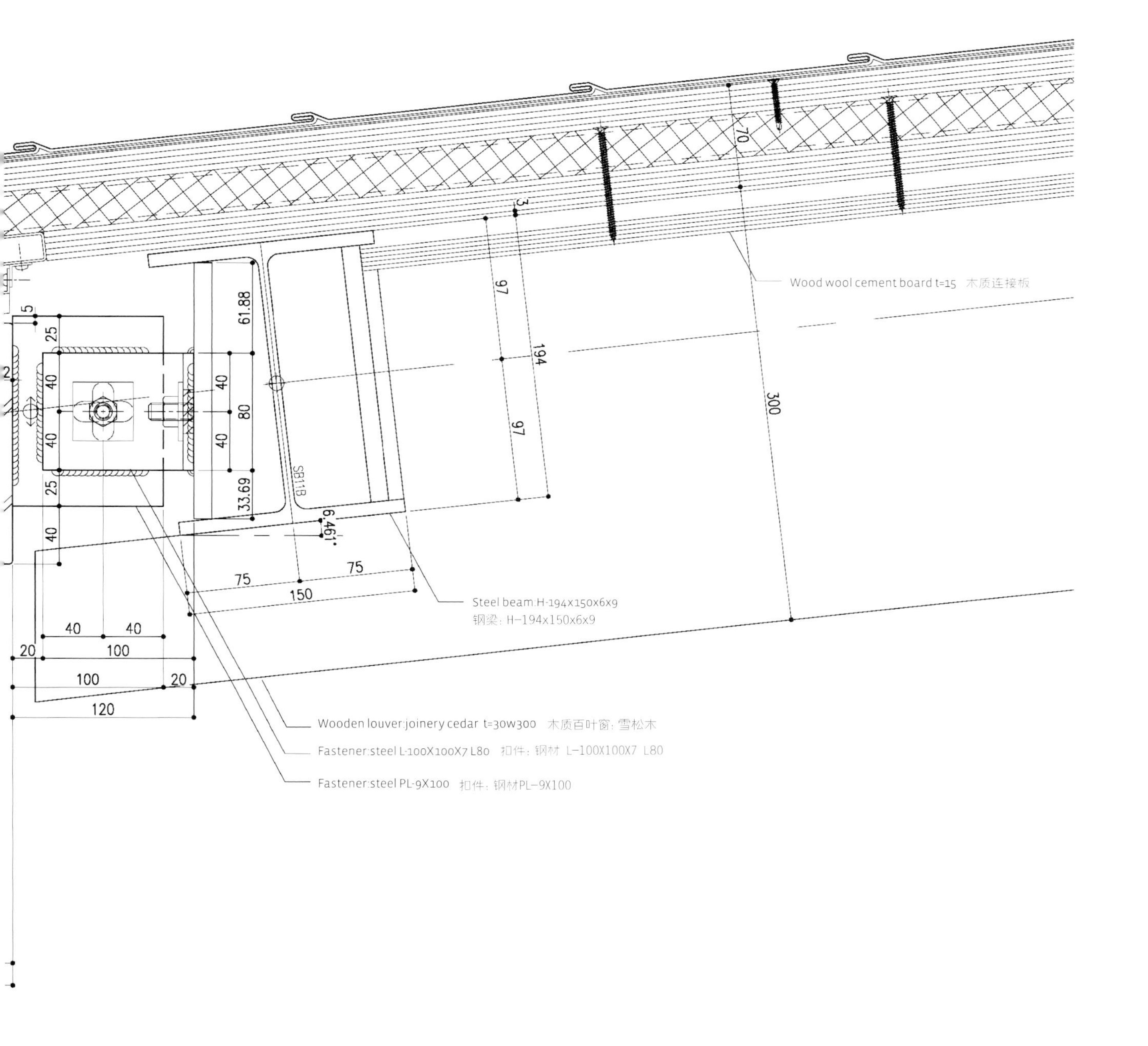
Wood wool cement board t=15 木质连接板
Steel beam:H-194x150x6x9
钢梁：H-194x150x6x9
Wooden louver:joinery cedar t=30w300 木质百叶窗：雪松木
Fastener:steel L-100X100X7 L80 扣件：钢材 L-100X100X7 L80
Fastener:steel PL-9X100 扣件：钢材PL-9X100
SB11B

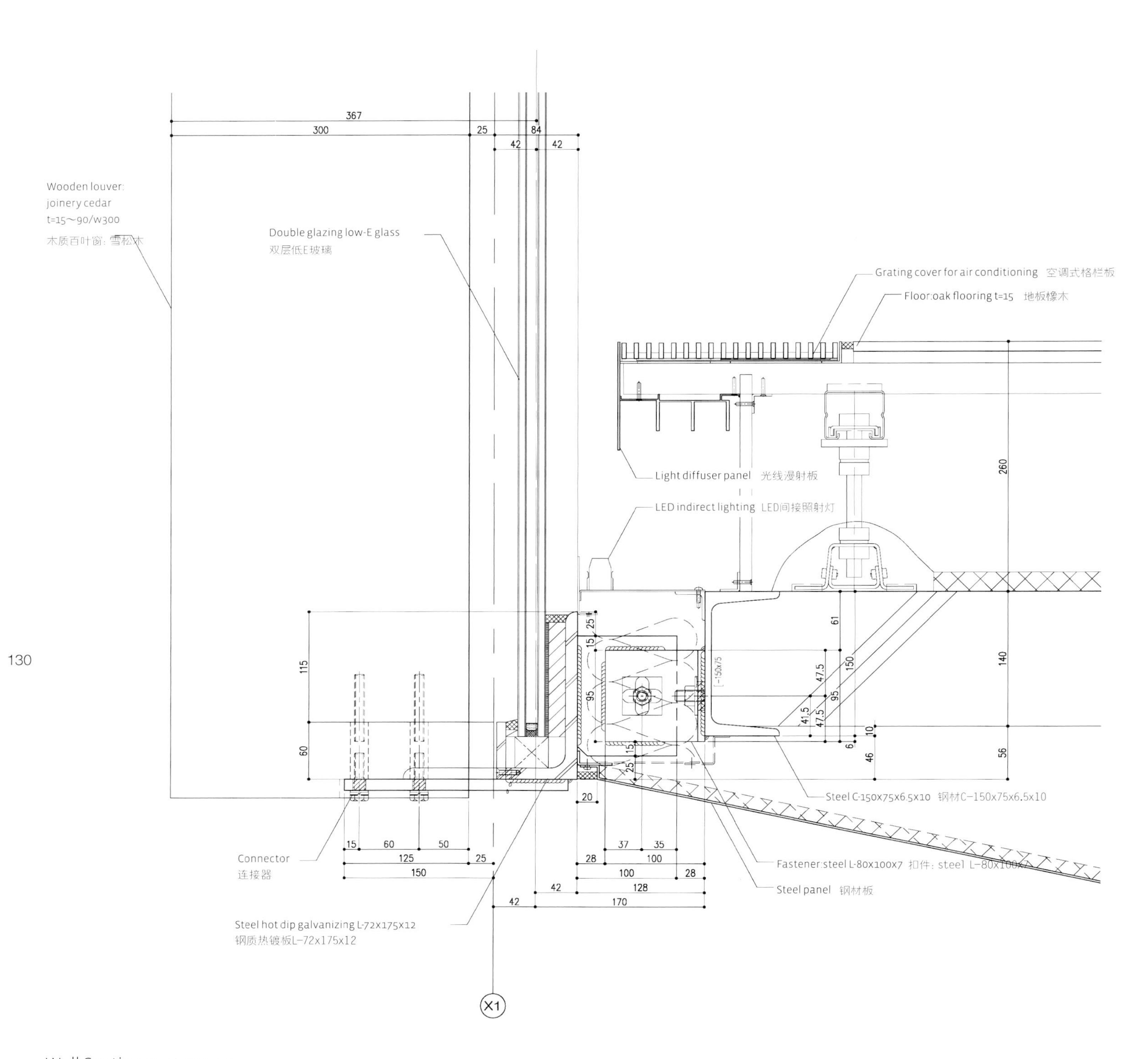

Wall Section 2 外墙剖面2

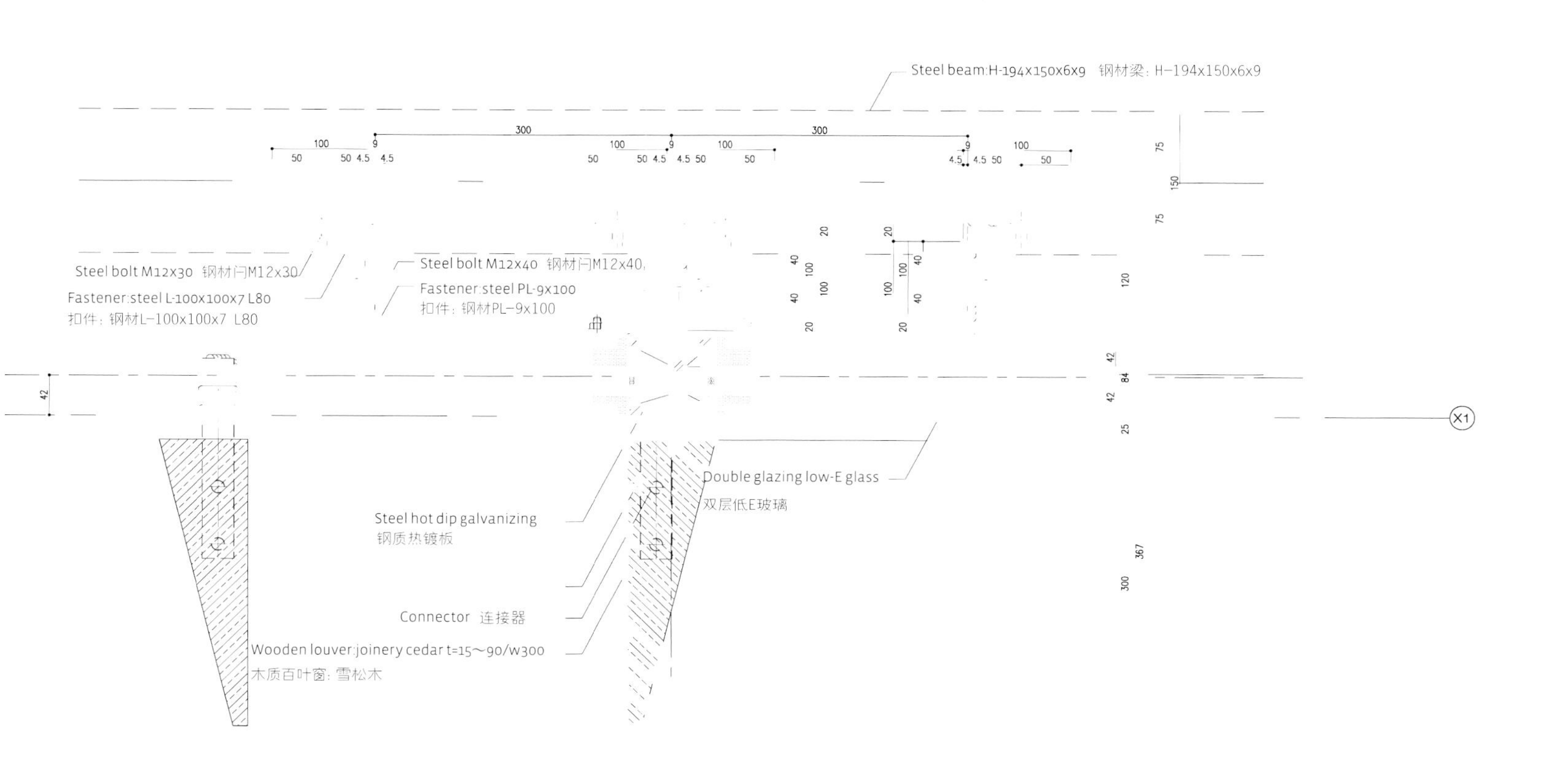

Wall Section Plan

外墙剖面平面图

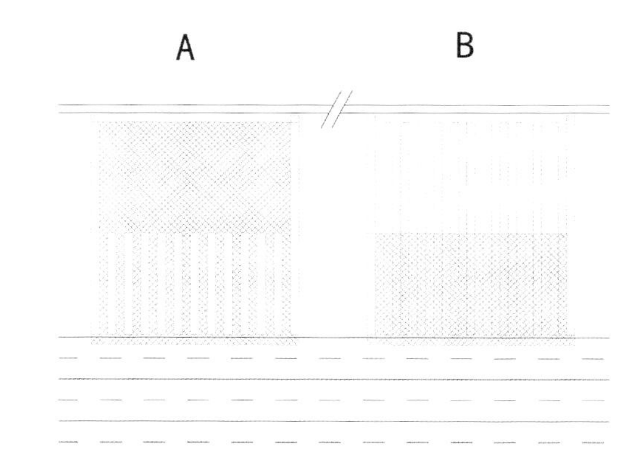

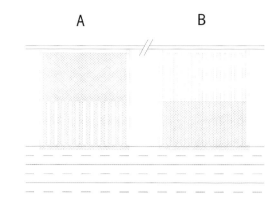

Handrail Details 扶手细部

East Elevation
东立面

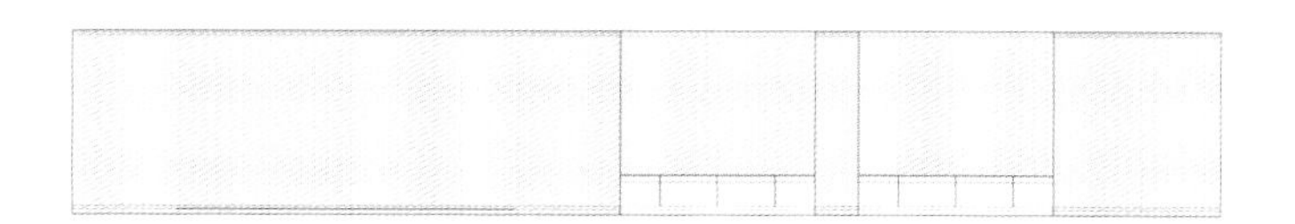

West Elevation
西立面

The courtyard made up of bamboo surfaces inherits the garden elements and blends into nature. Wandering in the courtyard, you'll find bamboo is interwoven and creates a crossing visual effect. The tall bamboo encloses an outdoor pedestrian walkway and forms the dissymmetrical layout. From the outer appearance, the courtyard is the false or true square. With illumination, its vertical lines are clearer. Its simple appearance interprets the unity between architecture and nature.

由竹子外表皮所构成的院子，沿袭中国传统园林的基本元素，融入自然的环境。漫步院中，竹子纵横交错，营造纵向横向视觉效果。高高的竹条围合成户外步行道，在湖面上呈不对称布局。从外观上看，竹院是一个有虚实变化的立方体。配合灯光的照射，其竖向线条更加明显。简洁的外形诠释了建筑与自然的统一。

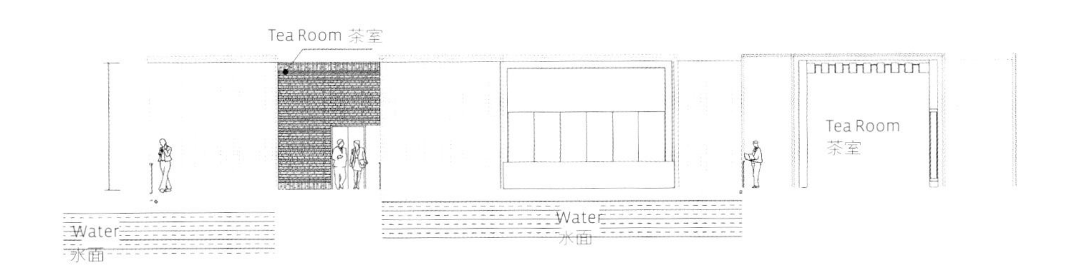

Section A-A

剖面图

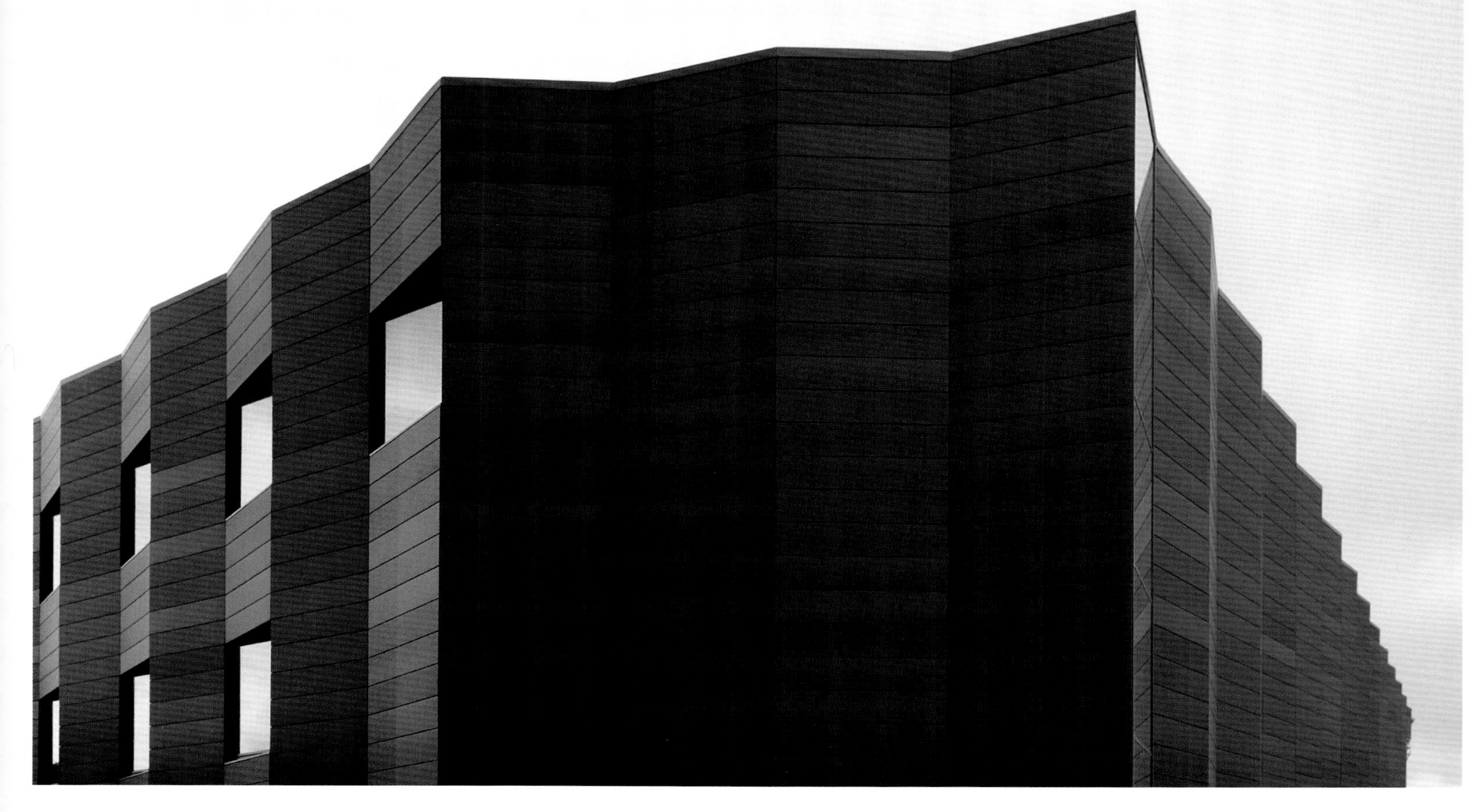

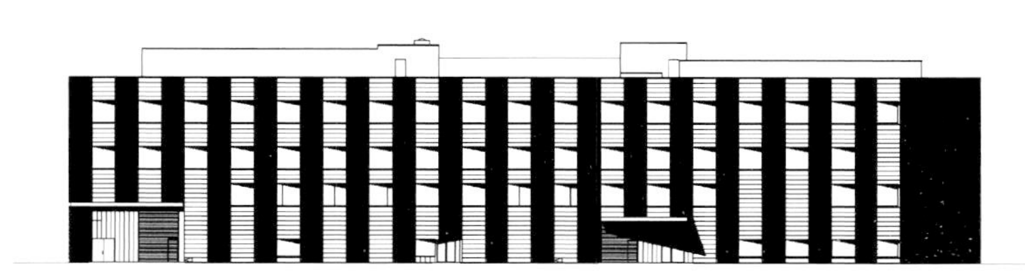

East Elevation

东立面

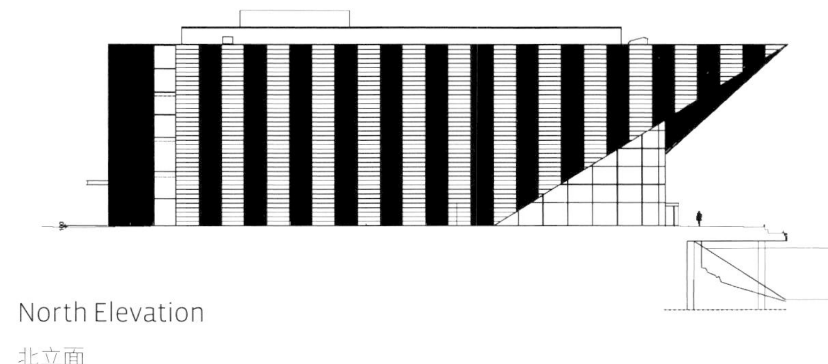

North Elevation

北立面

The façade is made of wave curves and the highlight is that one side of it is covered by the wave wood wall which is made of local oak whose hung structure like a huge tree crown flexibly distinguish reality from fantasy. Walking through the building, you will feel the alternation of natural landscape and art. The rest of the façade is composed vertical lines to give buildings an elegant form. The whole building is clearly defined and its clarity is exaggerated and it stands firmly above the awkward industrial surroundings.

整个建筑表皮由波浪形的曲线构成，最突出的地方是其正面被一面波浪形状的木墙包围。这种抽象的纪念碑式墙体由当地的橡树制成，悬挑的木结构，就像一个巨大的树冠，巧妙地将现实与幻想区分开。如果穿梭其中，会感受到天然景观与艺术的交替。其他立面由垂直折叠表面构成，赋予建筑物一种柔和而优雅的外形。整座建筑轮廓分明，表现之清晰几乎有点夸张，高傲地屹立在周围恶劣的工业环境之中。

To sustain the local industrial tradition, the building adopts the toughened glass, with the rotten red lattice shutters opening and closing, creating a vivid façade image.

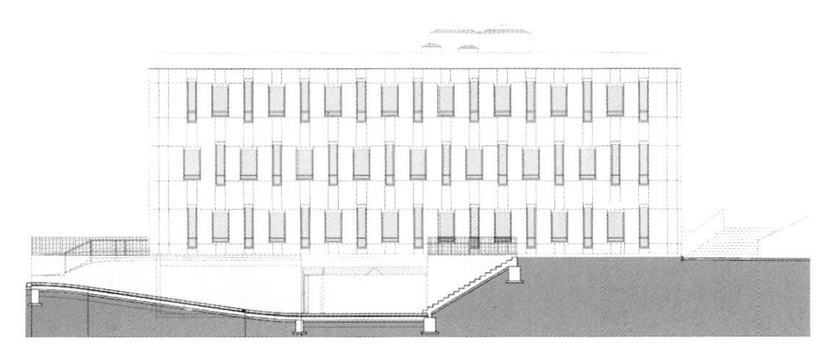

East Elevation 东立面

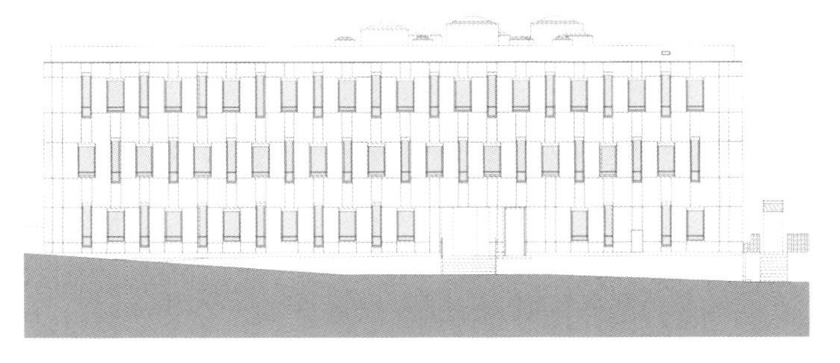

South Elevation 南立面

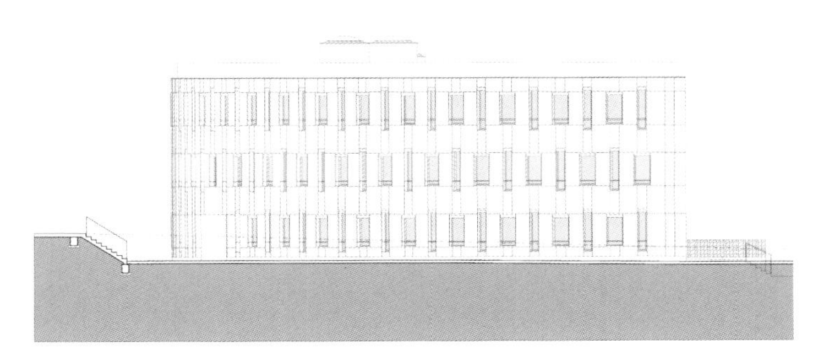

West Elevation 西立面

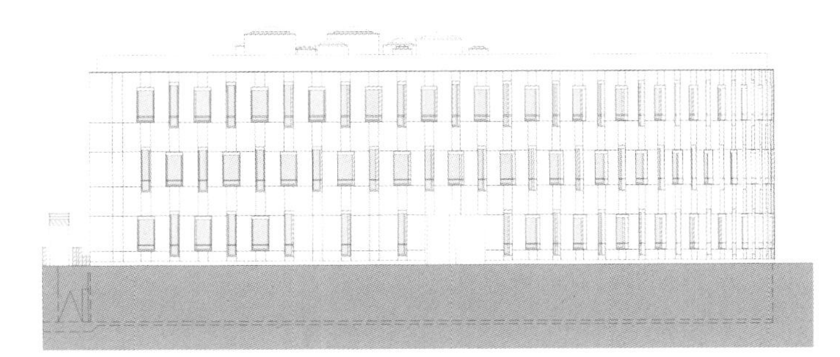

North Elevation 北立面

为了延续该地工业化的传统，建筑采用了耐候钢化表面，生锈的红色网格通过百叶窗的开放和关闭，创造了充满活力的表皮形象。

The winding vertical firs become the spot scenery of the tunnel entrance. The rectangular section of firs in different heights applies parametric design to display its complexity. One ruled surface motion reflects the visual effect of topography restraints and cinema grades. Viewed in different positions, it is sometimes visible and sometimes opaque, and forms an environmental sculpture with striking city features.

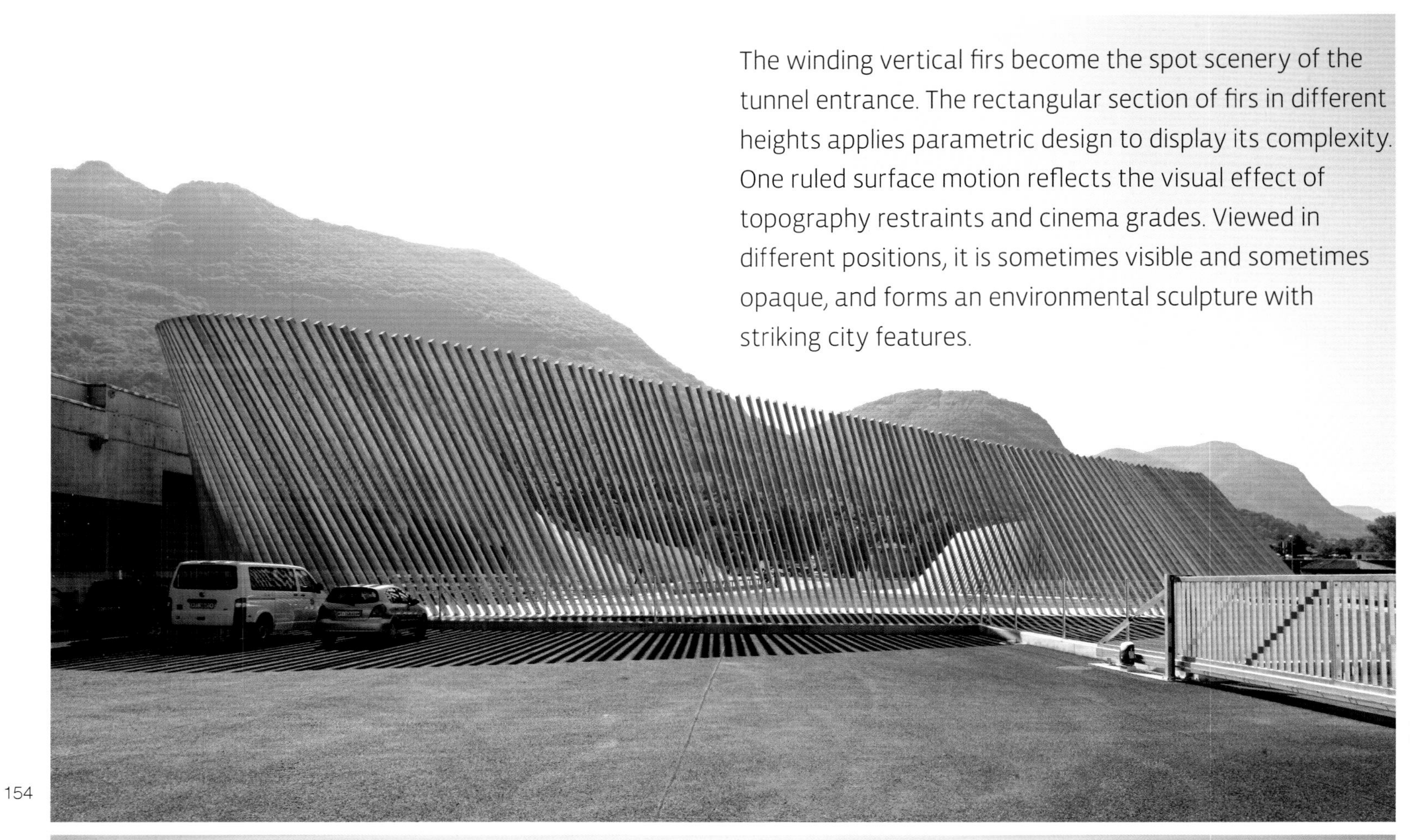

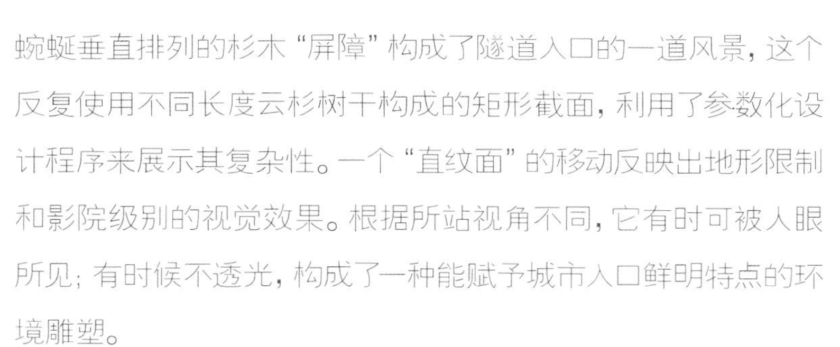

蜿蜒垂直排列的杉木“屏障”构成了隧道入口的一道风景，这个反复使用不同长度云杉树干构成的矩形截面，利用了参数化设计程序来展示其复杂性。一个“直纹面”的移动反映出地形限制和影院级别的视觉效果。根据所站视角不同，它有时可被人眼所见；有时候不透光，构成了一种能赋予城市入口鲜明特点的环境雕塑。

Broken Line – Compared with the linear façade, the façade of broken line is gentler and more elegant, with an irritating character and rational rhythm. The inflection point of broken lines is powerful as can be seen in the architecture character and it breaks the dullness of traditional buildings. The vertical broken line is very lively and makes the building see to be tall, dynamic and vigorous and brings vigor to the city.

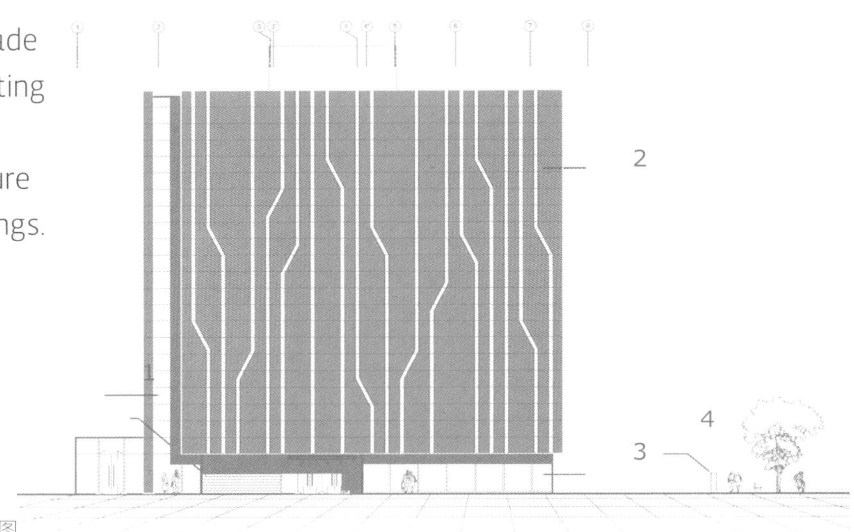

Front Elevation 立面图

1. Cristal Arenado 水晶
2. Cristal Negro Con Detalles En Cristal Blanco 水晶黑与白水晶细节
3. Cristal Arenado 水晶
4. Poste Luminoso 杆腔

折线条构图——折线形立面与直线立面相比显得更温和、优雅，又具有不安定的性格，它表现出一种感性与理性相融的节奏。折线是由直线曲折改变方向而形成，其拐点牵系着两端的线段，显得聚集力强，使建筑呈现出刚锐、跃动、激发的品性，打破了传统的规整与沉闷。在建筑中所运用的竖向的折线，给人规整中又不失活泼的感觉。由于竖直的折线的透视汇聚，会使建筑物显得更加高耸而有活力，呈现动态的美感，给建筑和城市带来生机。

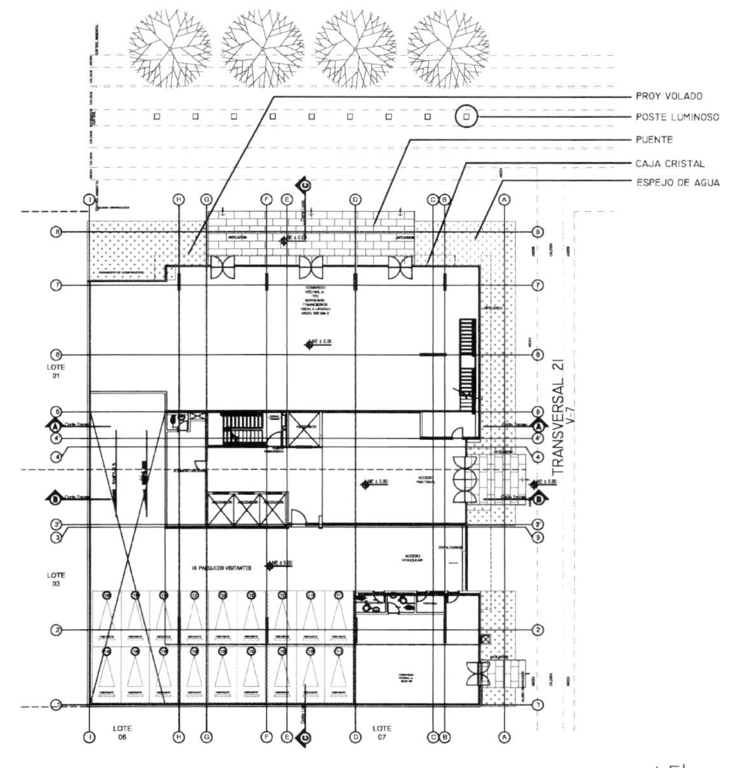

1st Floor 一层平面图

The flowing straight line planes, used in the architecture surface are arranged in a fixed way and modified slightly in the organized chart, with a curve-like feeling. It is a combination of both regular and irregular facades which will thus be more creative, more fashionable and more vigorous.

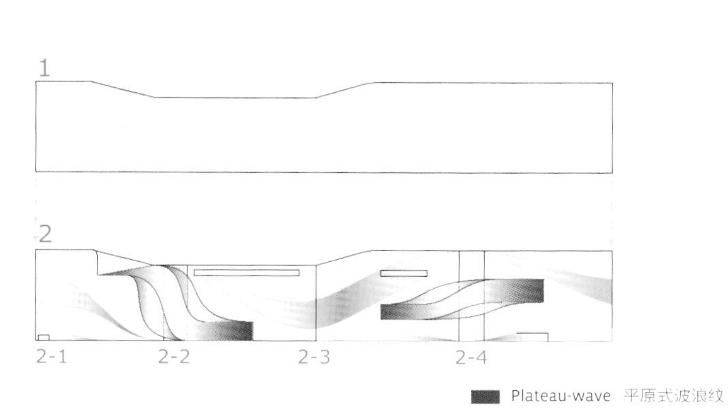

Plateau-wave 平原式波浪纹

Generated subwave 自带式次波纹

2 set optical illusion 两套装置造成错觉

3

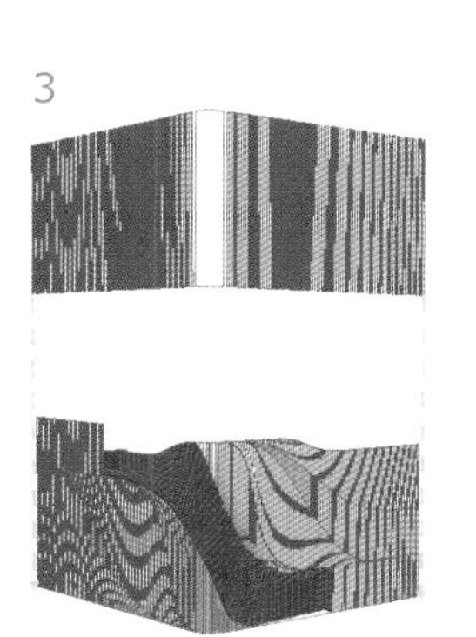

4

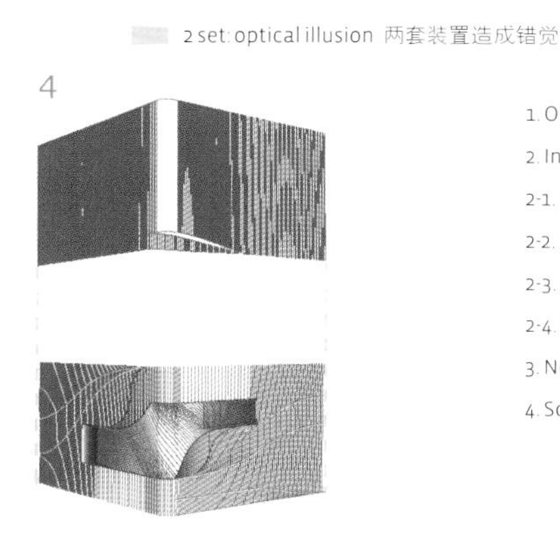

1. Outer layer 外层
2. Inner layer 内层
2-1. West 西面
2-2. South 南面
2-3. East 东面
2-4. North 北面
3. North-west 西北
4. South-east 东南

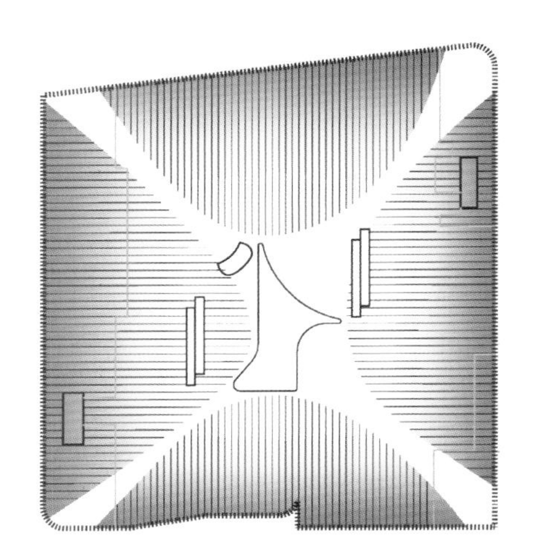

"Outside in" feature from scenario 1

步骤1中嵌入

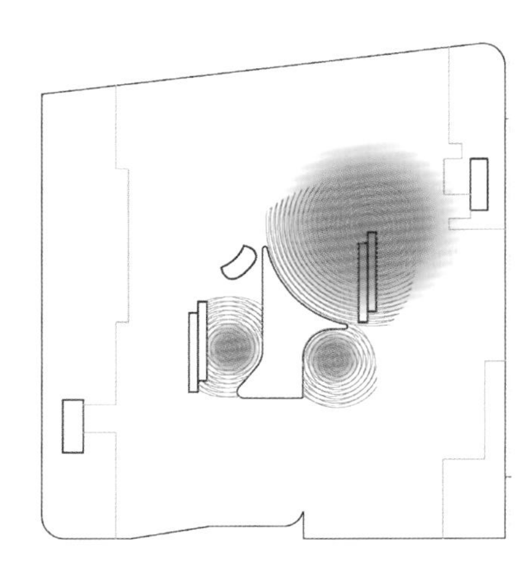

"inside out" feature of scenario 2

步骤2中

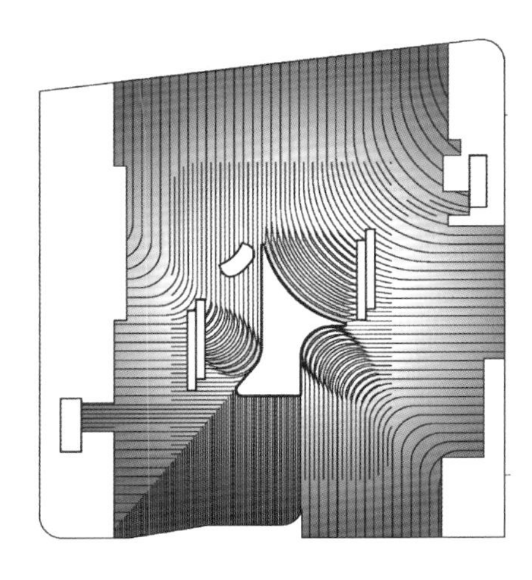

Final systesis

最后步骤

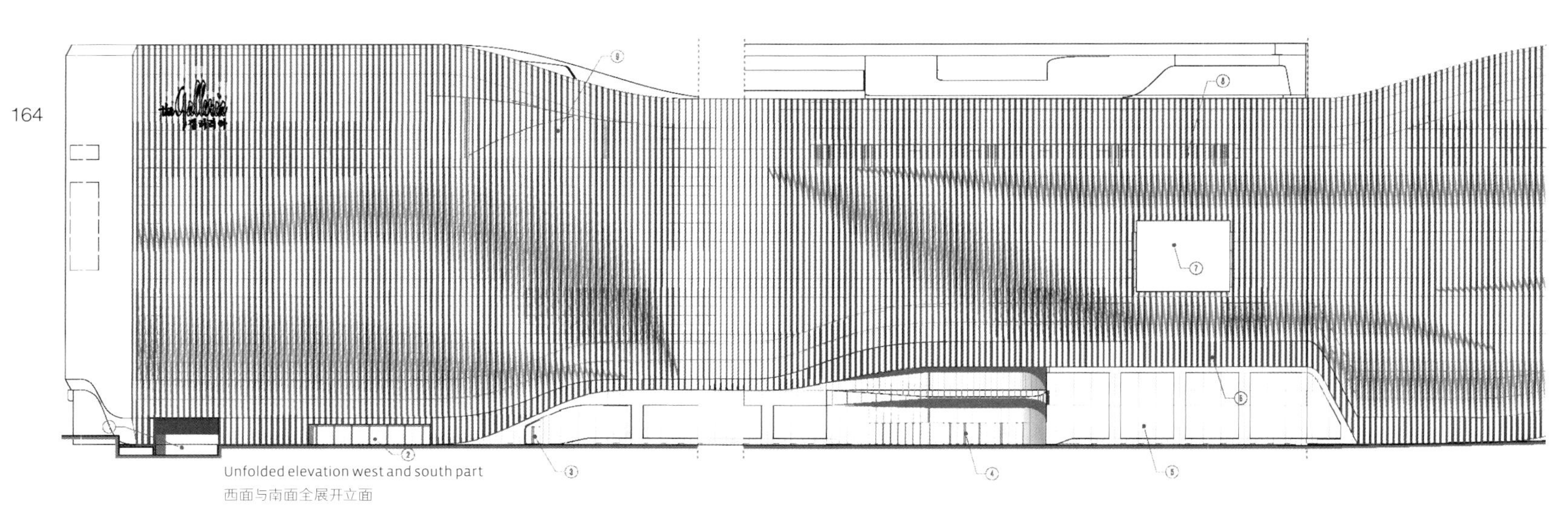

Unfolded elevation west and south part

西面与南面全展开立面

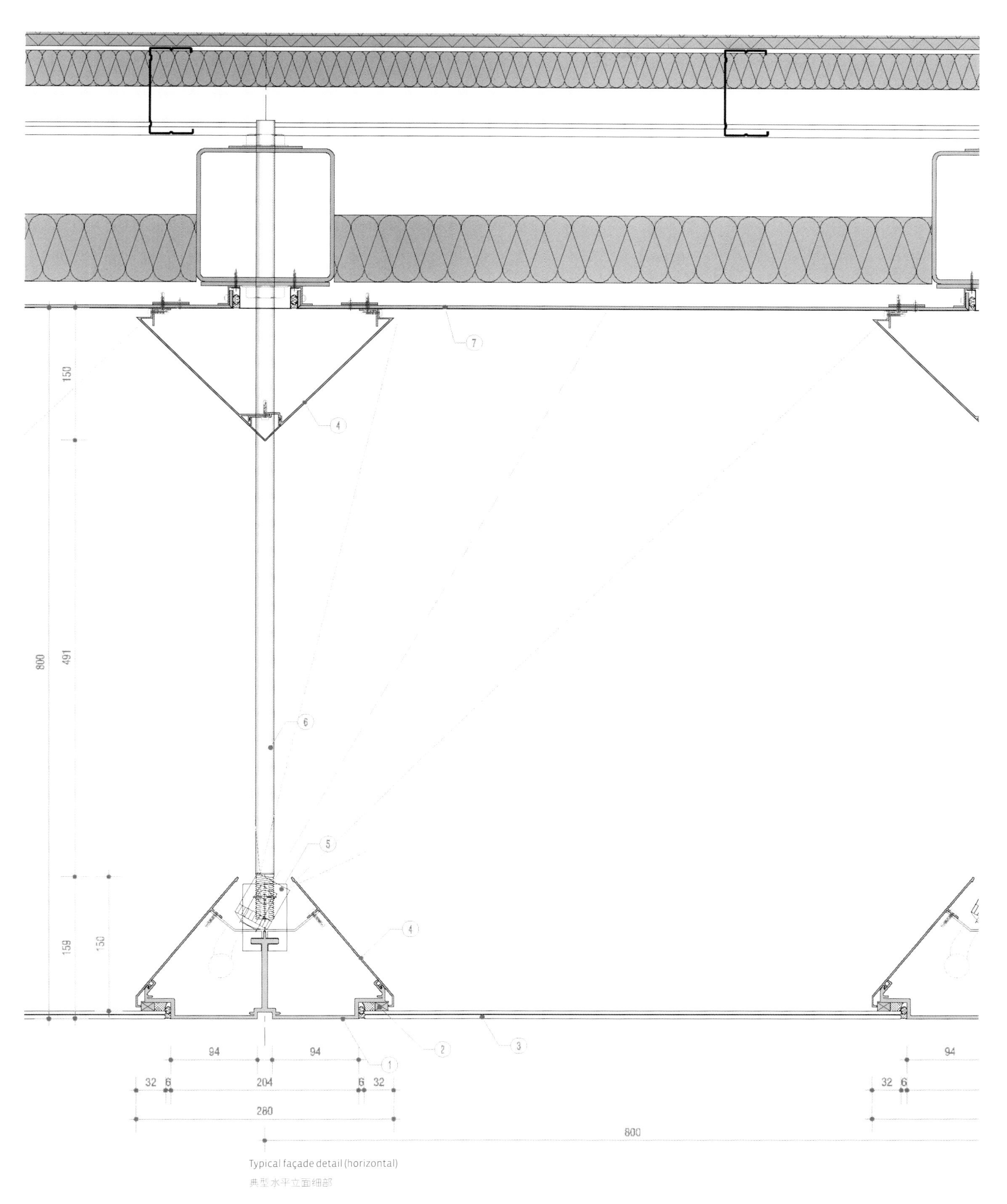

Typical façade detail (horizontal)
典型水平立面细部

SALE
SALE

这种流动的直线平面运用于建筑表面上，是按照某种固定的形式进行线的组合，在组合图形中又加以局部变化，产生曲线的感觉，具有流动感，是建筑立面规则与不规则的组合方式，可以使立面构成变得丰富而有创意，予人以时尚与活力。

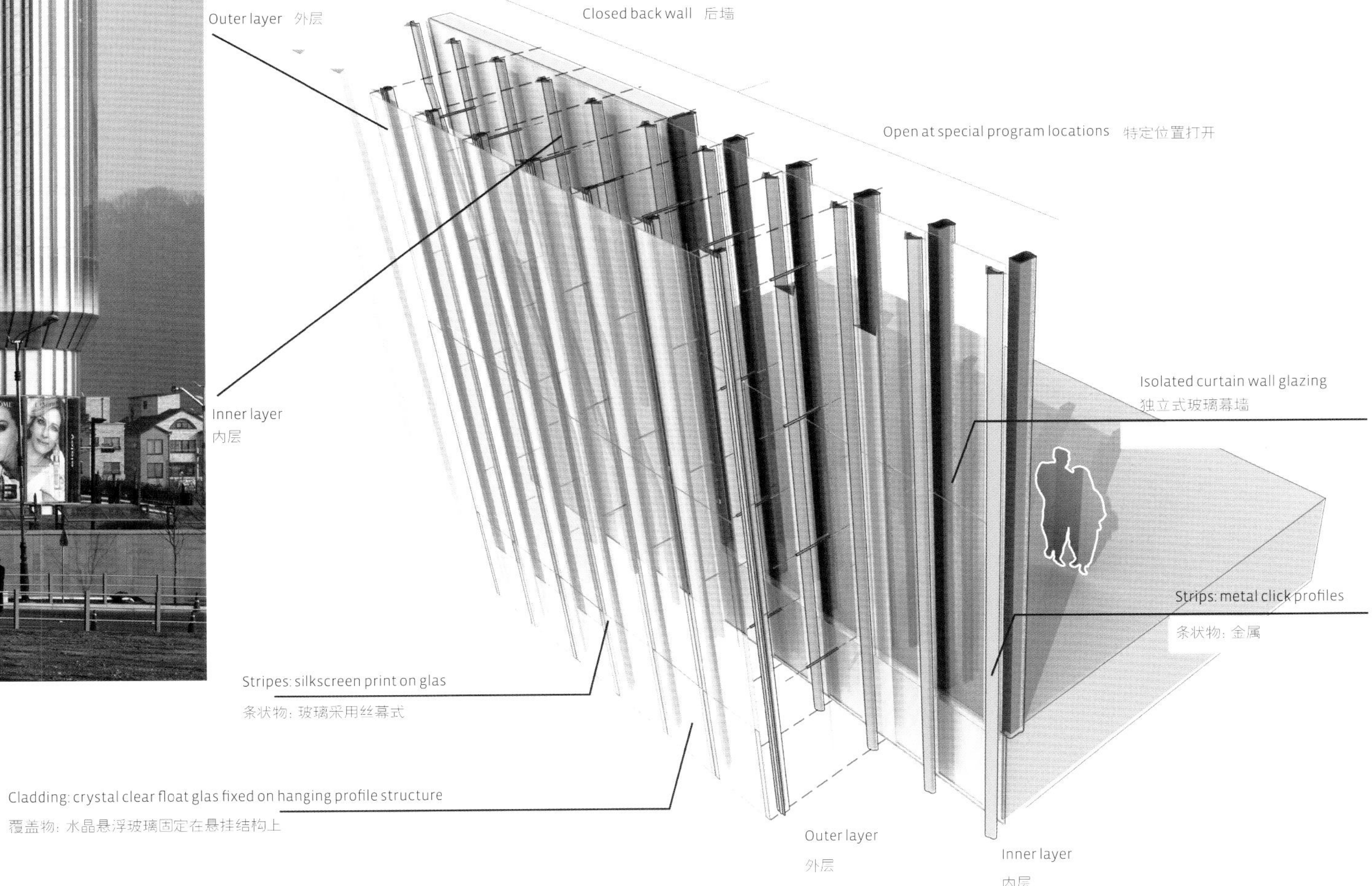

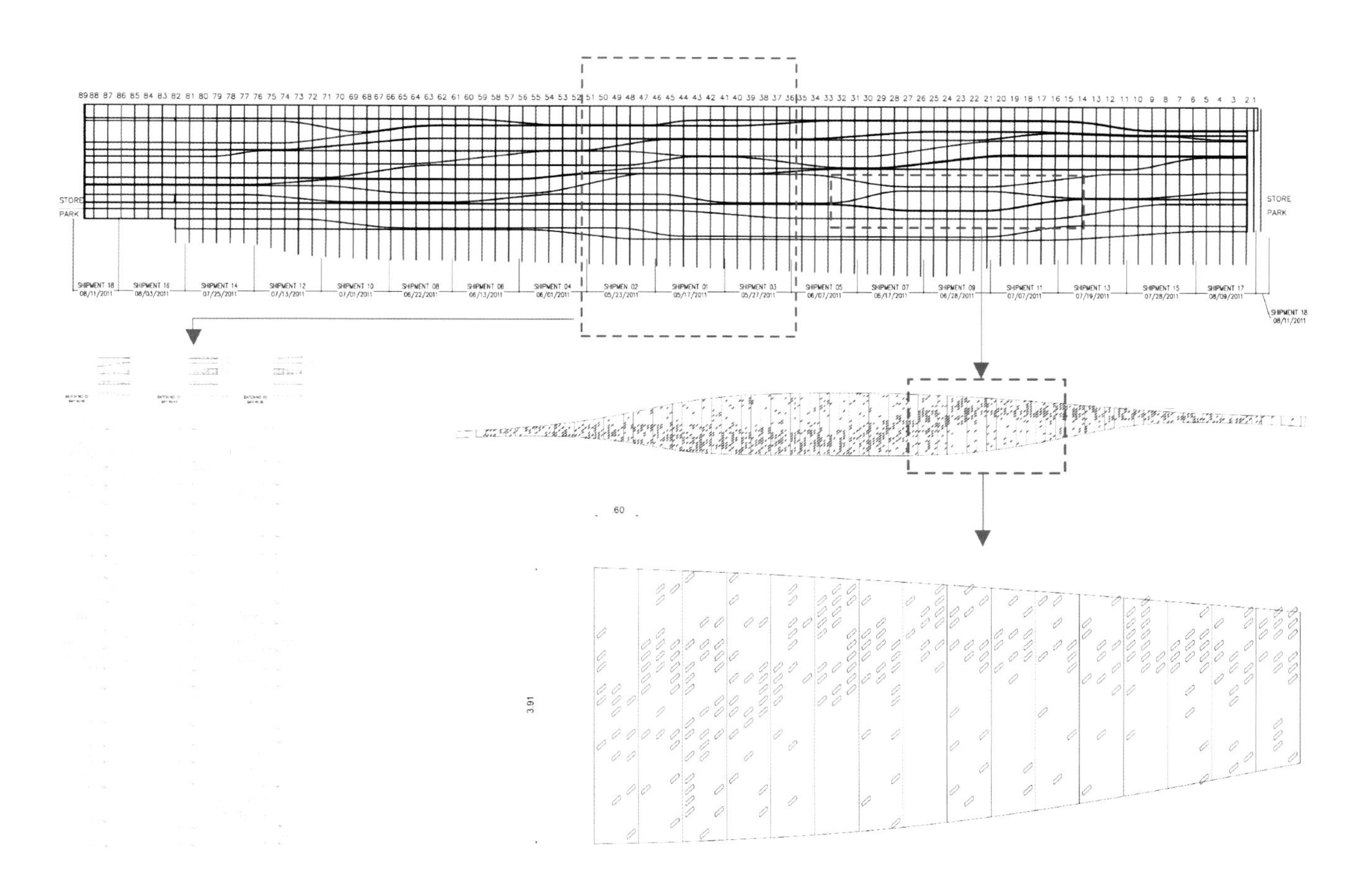

The building aims to create a vivid image and its façade stands firmly at the crossing. By means of 3-D software, the designer designs the powerful, flexible and fluid architectural surfaces. The smooth double-layer curtain wall separates the messy surroundings with the inside. The smooth stainless steel surface becomes part of the city. At night, the wrinkles on the surface become a subtle landscape.

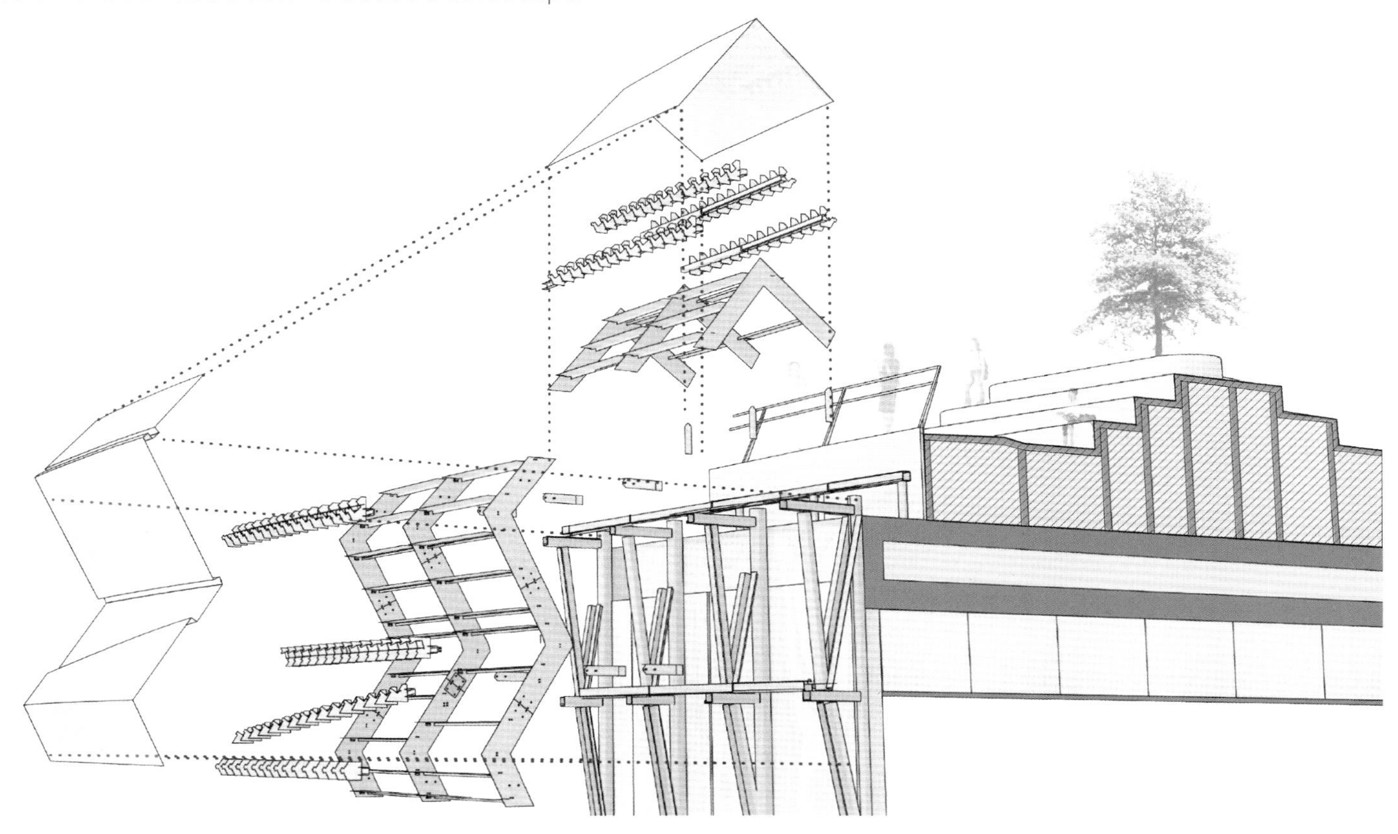

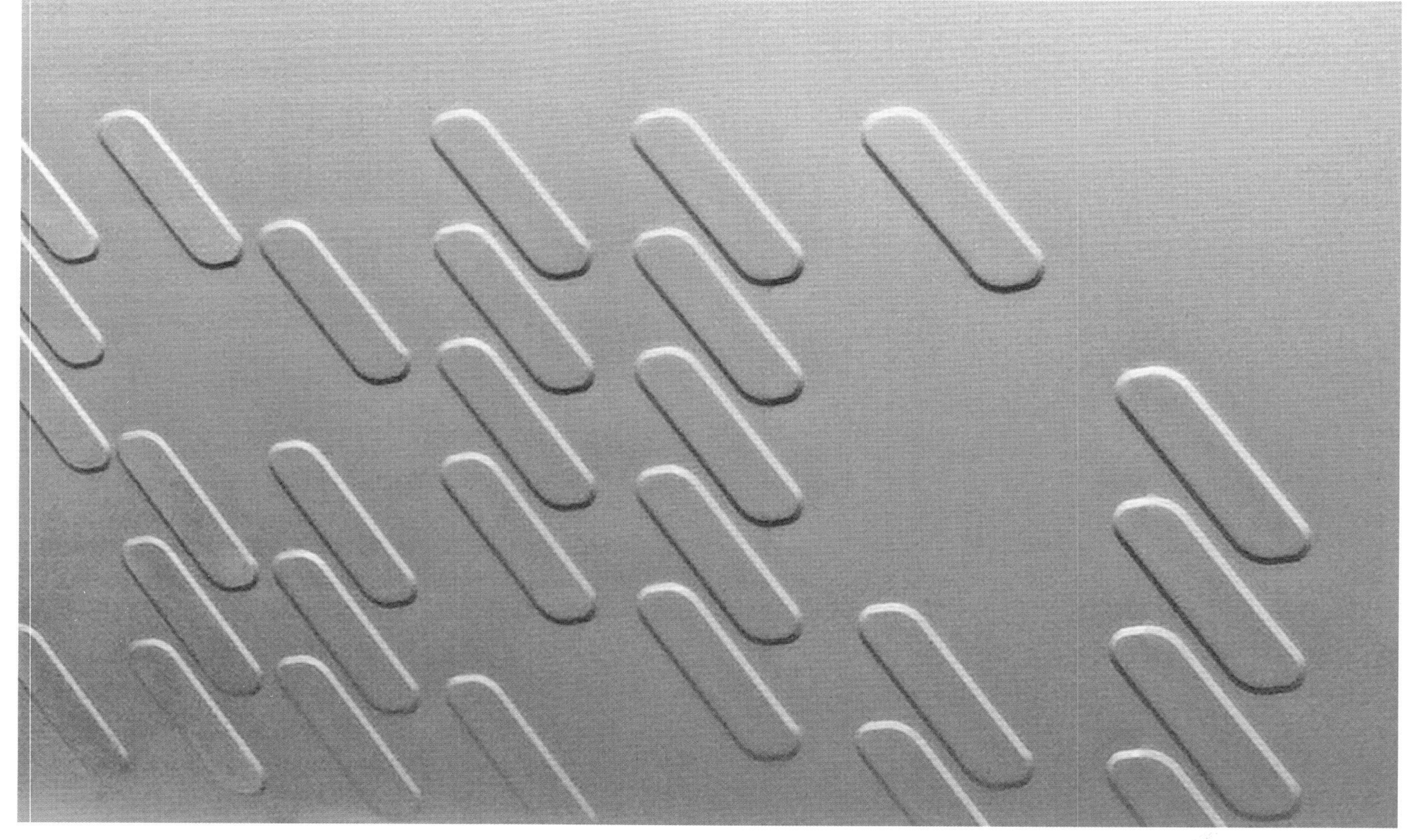

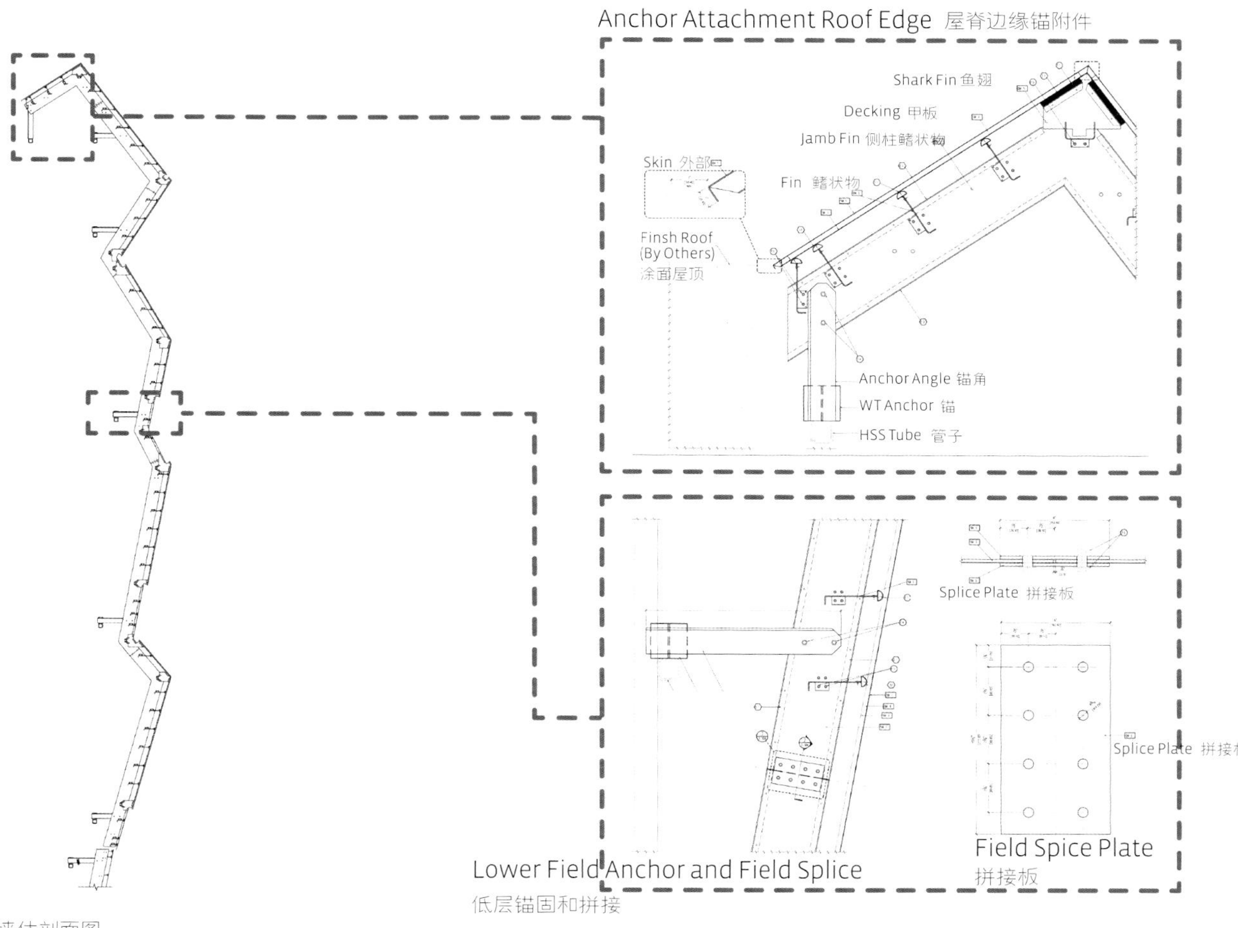

Wall Frame Section 墙体剖面图

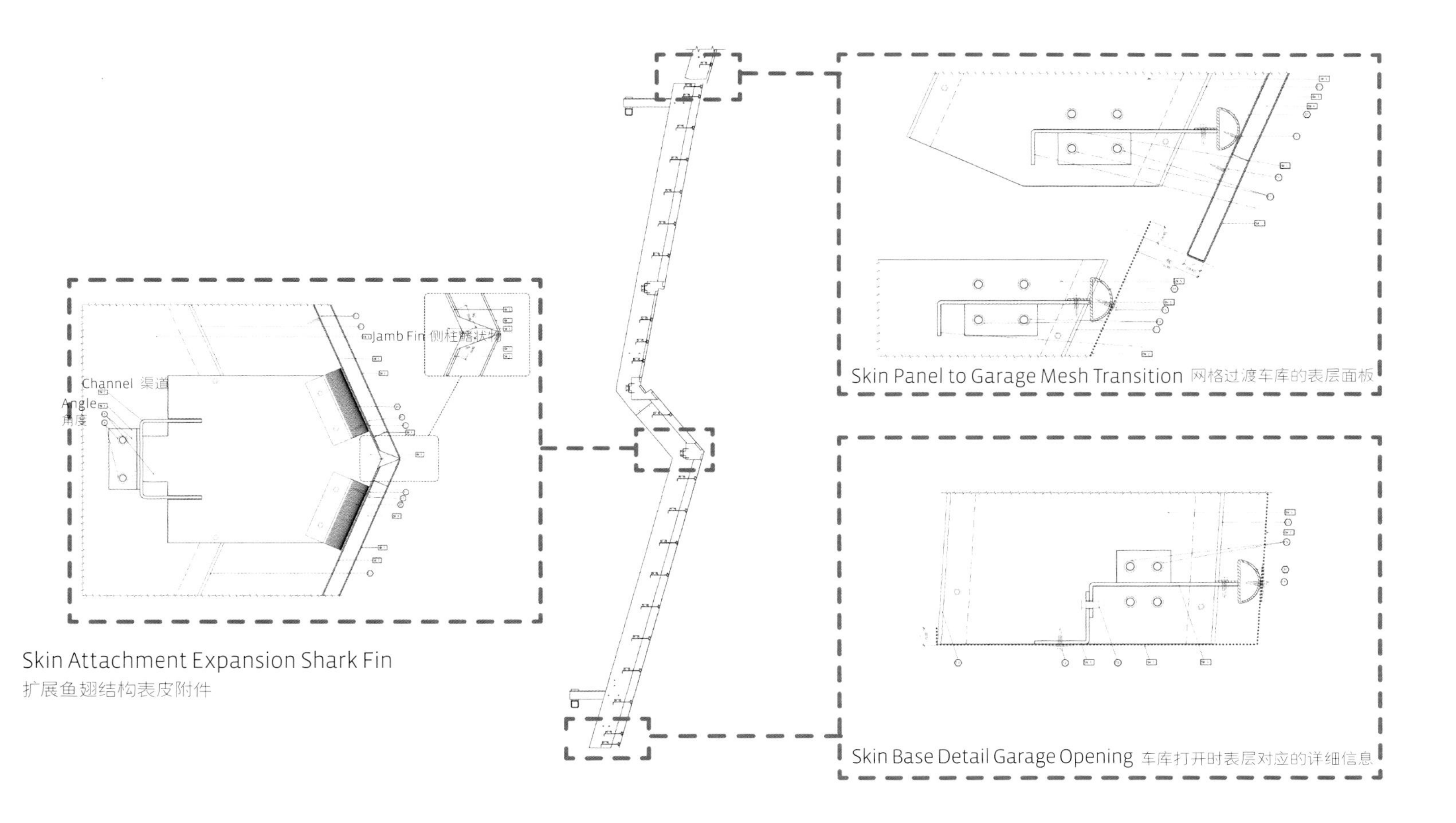

Skin Attachment Expansion Shark Fin
扩展鱼翅结构表皮附件

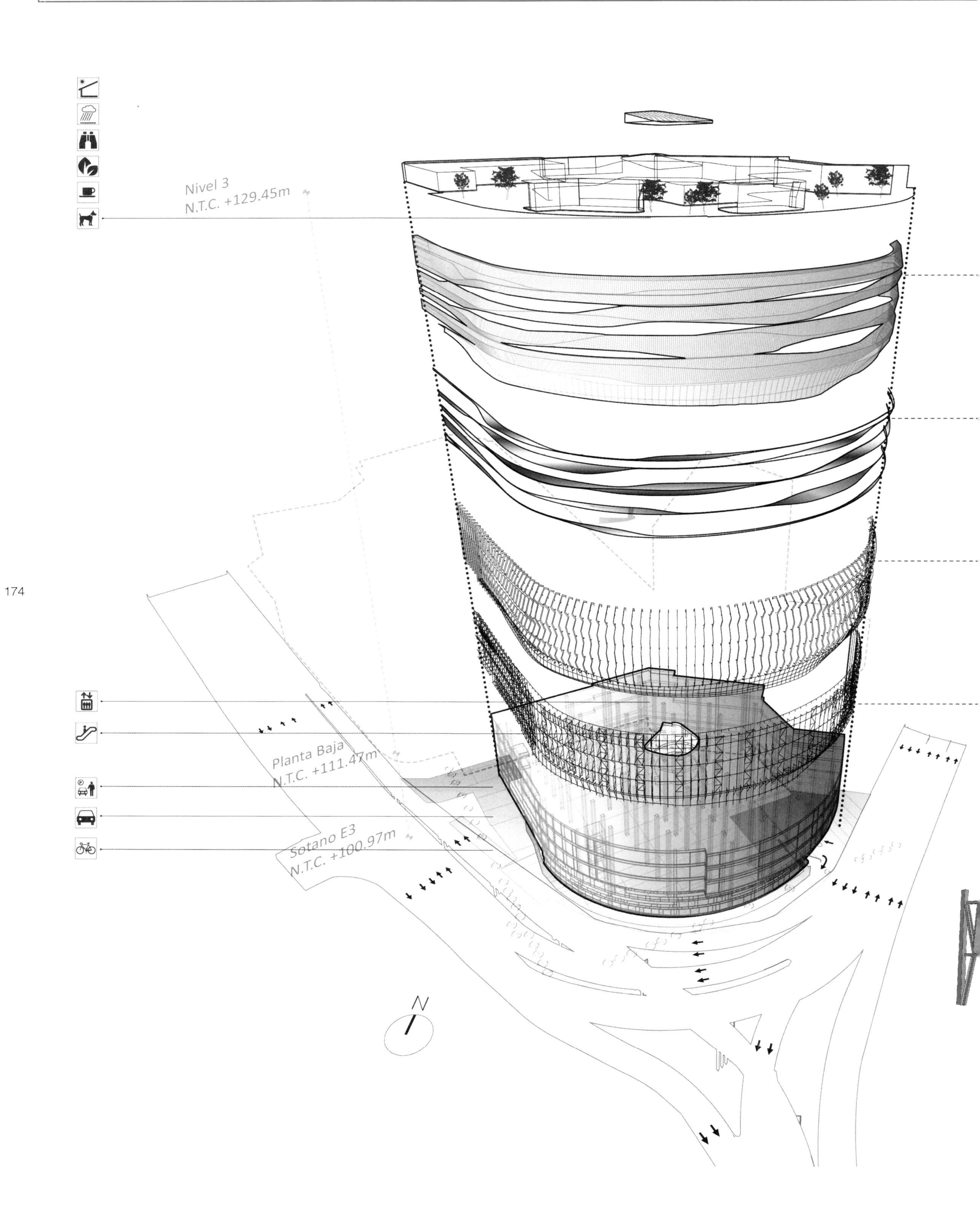
Nivel 3
N.T.C. +129.45m
Planta Baja
N.T.C. +111.47m
Sotano E3
N.T.C. +100.97m
N

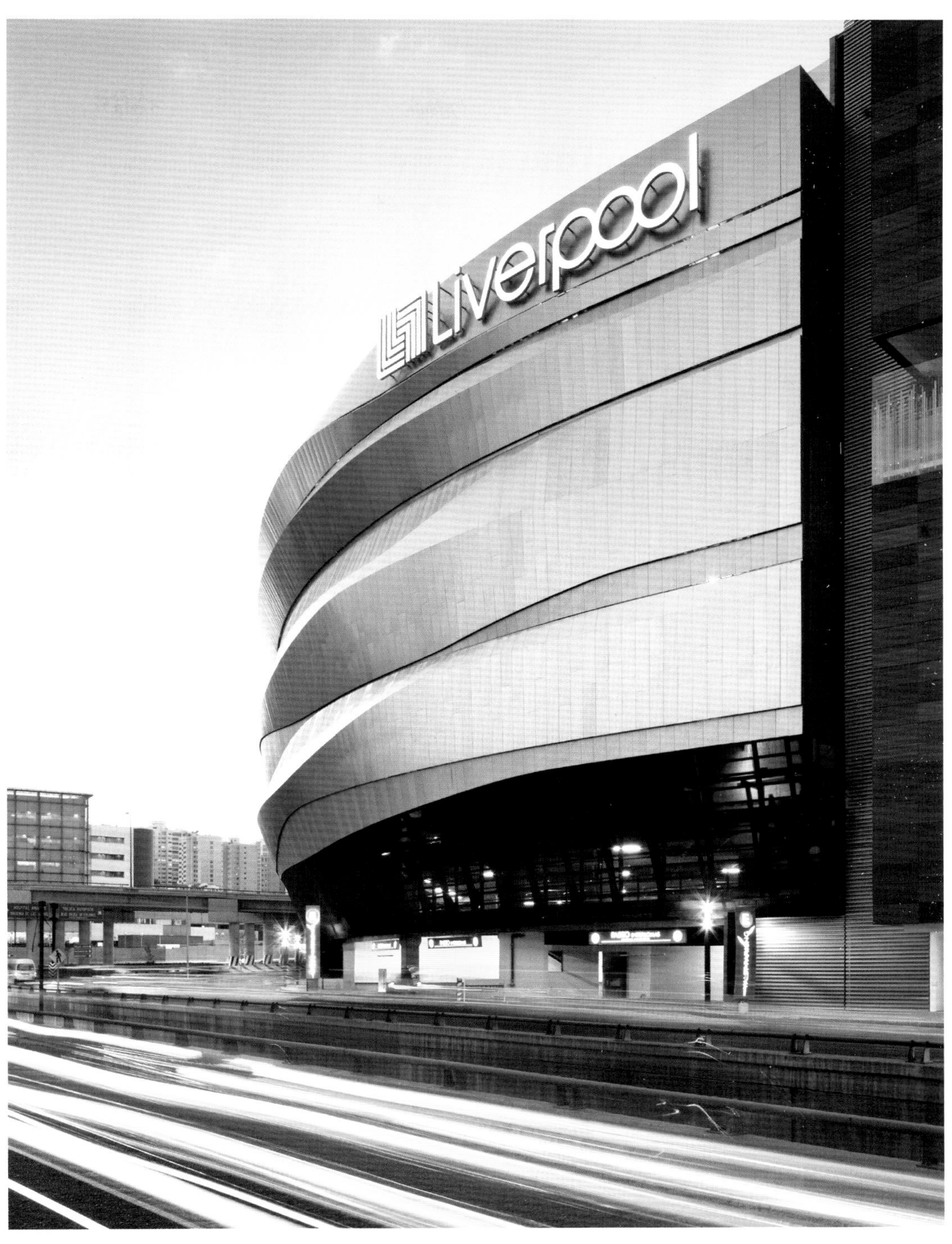

该项目旨在创造一个充满活力和现代感的新形象，建筑外表皮在这个布满立交桥的区域以刀锋般的锐气傲立于路口。设计师利用三维软件设计了这个有活力、灵活、流线型的建筑外观。流畅的双层幕墙将混乱的环境和内部隔开。光滑的不锈钢外皮沐浴在阳光中并成为这个城市的一部分。到了夜晚，光在外表皮的褶皱中形成微妙景观。

Palacio de Hierro

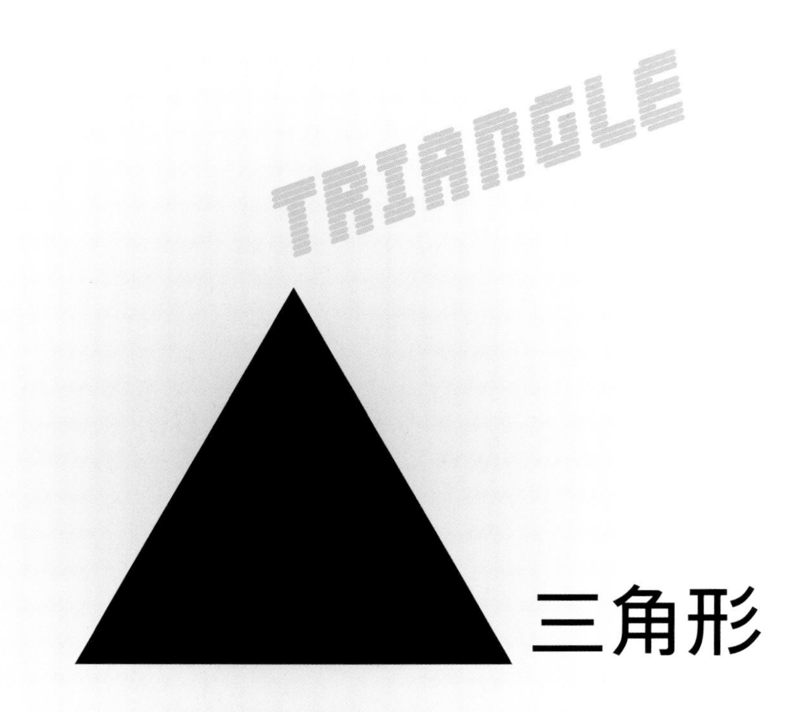

LEADING PART STANDING OUT, WITH A STRONG VISUAL SHOCK

主次分明，重点突出，具有强烈的视觉冲击效果

In nature, triangle is often seen in the highlands, valleys, plants, etc. The natural phenomenon reveals the nature of triangle. On one hand, it is a result of outside force requirement for stabilization, but on the other hand the structural character of triangles themselves. This is different from equilibrium of round and functionality of square.

The triangle façade is attractive in vision and it is very striking in the main part, just like that in the mountain top building. Meanwhile triangles are greatly suited to other geometries and can direct people's attention to the top position. Triangle itself is a concept of figure and but in architecture it is not just a planar figure and people can have different understands of it. It can be the triangle plane, façade, triangle-like area, triangle decoration, etc. It gives people an enterprising spirit.

In architecture conception, triangle can be plane, space or flowing lines, and can also be structure, façade, body or unseen structure relations. In all planes, the triangle is the most unique. It is the basic geometry with the simplest components and conditions. For example, the square can be seen as the combination of two triangles while the circle is a result of indefinite triangles rotating around the center. From the perspective of geometry feature, a triangle is basic and pure.

在有形的自然界中，三角形态最为突出，无论高山峡谷还是树木花草，都显现典型的三角符号特征。这种自然意象的呈现暗示了三角形的自然属性，即一方面可以把它归于外界引力的结果，即力的稳定性需要而至的渐变，另一方面来自三角形本身内在稳定的结构特性。这明显区别于圆的圆融与均衡特征和方形的功能主义。

三角形立面具有突出的吸引视觉的能力，其构图容易做到主次分明，重点突出，呼应有序。如山顶建筑、高处事物等，同时三角形平面对几何图形有特殊的适应力，形成有冲击力的视觉效果，能够把人们的视线引向高处事物。三角形本身是一种形的概念，但它在建筑学中并不仅仅是平面图形，在建筑中人们对它可以有多种理解、感受和表达方式。它既可以是三角形的平面、三角形的立面、近似三角形的区域，或者是三角形的装饰……它带给人们的是一种积极向上的精神寄托。

在建筑设计构思中，三角形亦可以成为平面、空间、流线等等，还可以演绎为结构、立面、形体或者是看不见的结构功能关系。三角形在所有的平面图形中最具有其鲜明的独特性，它不仅是基本的平面几何形，而且是所有基本平面几何形中构成元素和条件最为简单的形。比如，所有四边形都可以看作是两个三角形的拼合，而圆形也可以认为是若干个三角形绕圆心排列的集合。从本质的几何属性来说，三角形本身所蕴含的几何性质更加的基本和纯粹。人们偏嗜于这种几何形态的运用，一方面是来自于三角形先天的力学优势，此外，它还有源于此衍生的独特的审美价值和哲学意味。

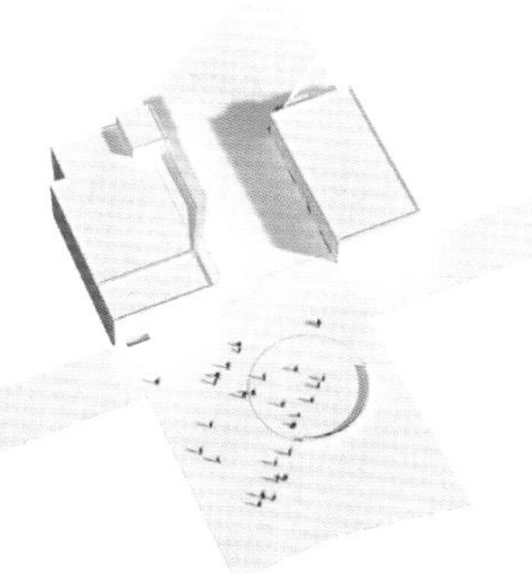

The existing site, between an old cinema and a cultural school, facing the town square

项目位于旧影院与学校间，面对城市广场

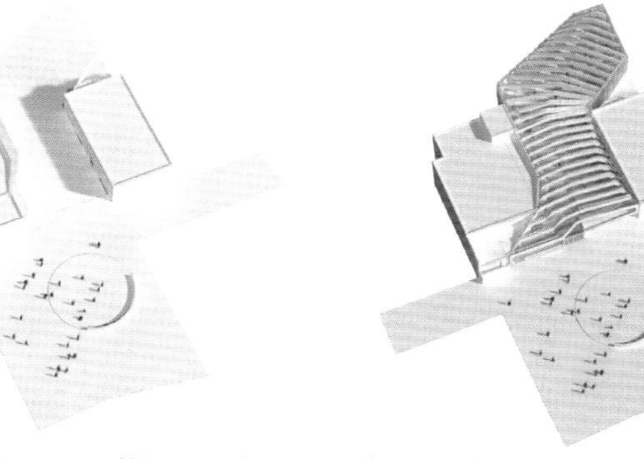

Rib structure in context

肋部结构

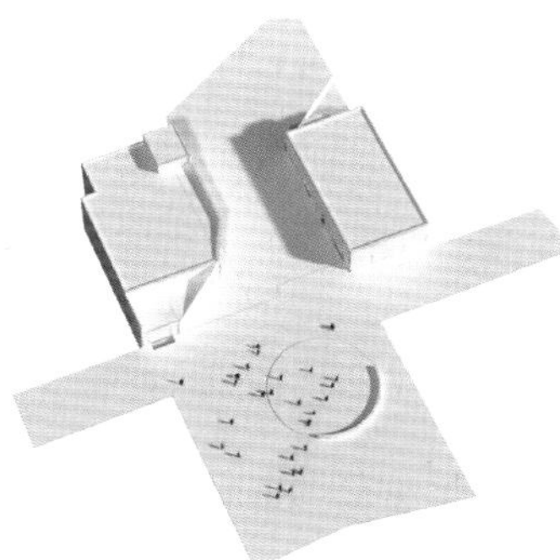

The existing cinema foyer is cut to maxim-ise the interface between the new building and the town aquare

电影院的大厅减少面积增加新建筑与城市广场的接触面

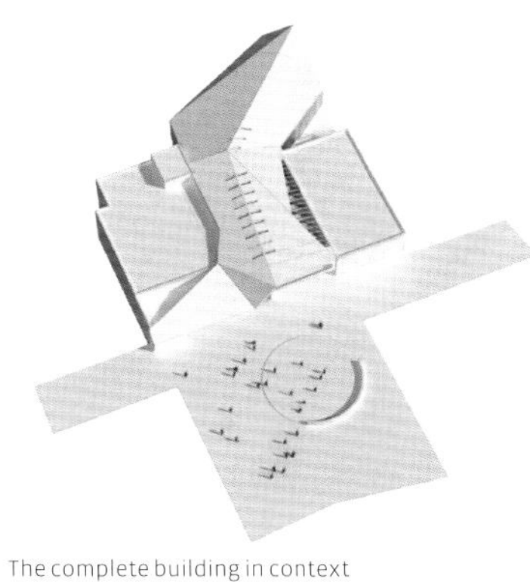

The complete building in context

完整建筑

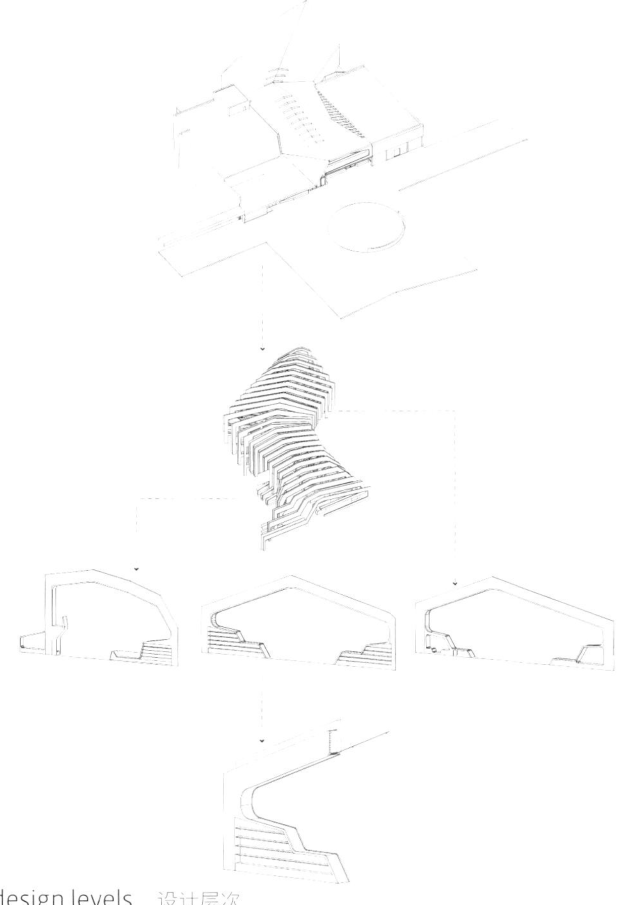

The design levels 设计层次

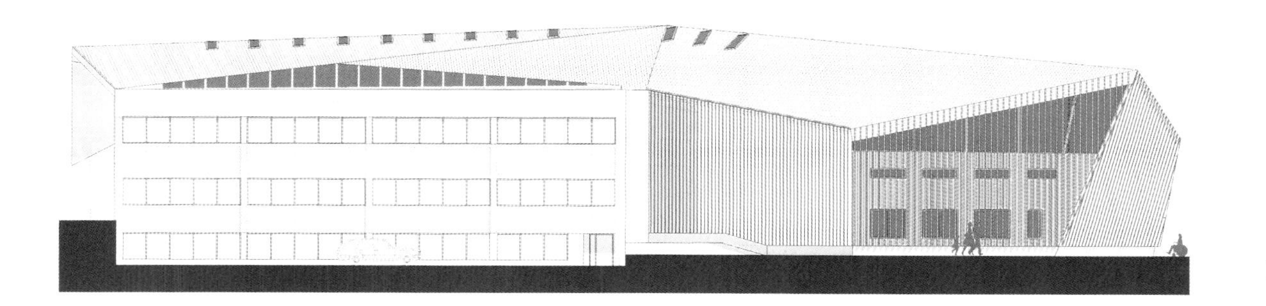

North elevation 北立面

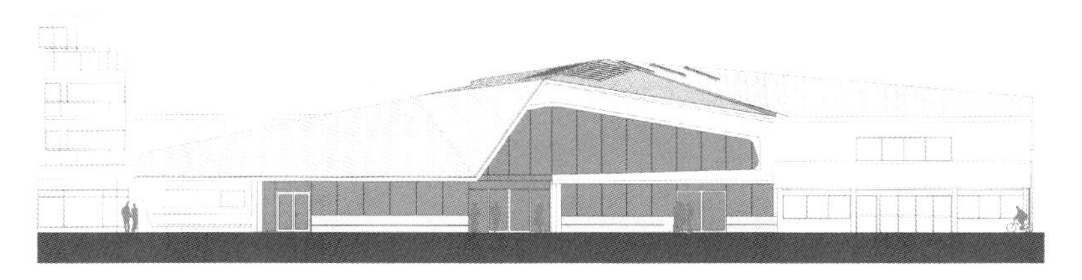

East elevation 东立面

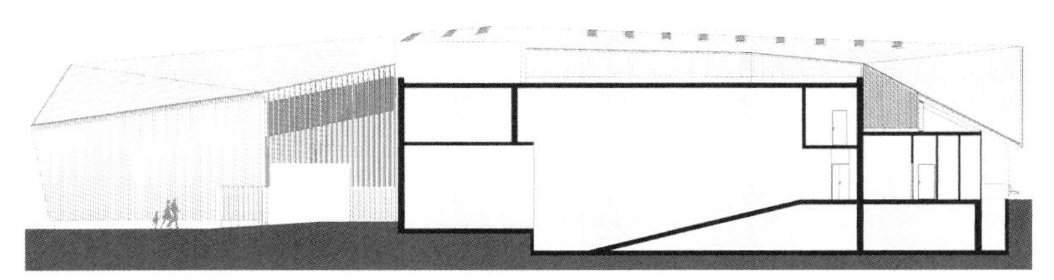

south elevation
南立面

The triangle library is very cordial in the city and the convenient access and the heights of surroundings building are very pleasant. There are in total 27 "ribs" used to express the geometry shape of the roof which unfolds openly to a natural transition, taking into consideration the surroundings and neighboring buildings. On the other side, the vertical sun shading board is installed on the exterior wall, providing gentle light to the inside space.

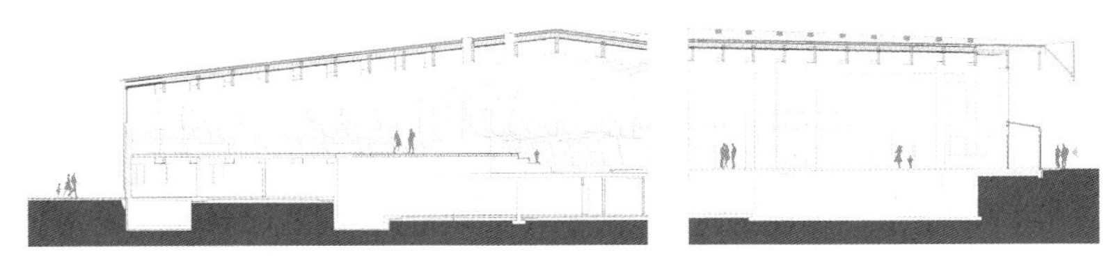

Long section 长剖面

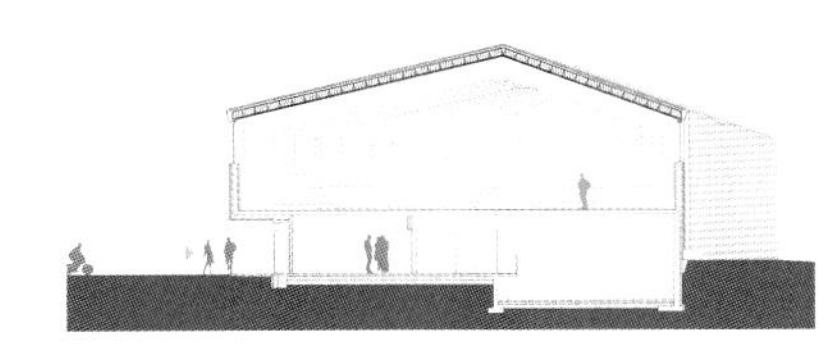

Cross section 1
横断剖面1

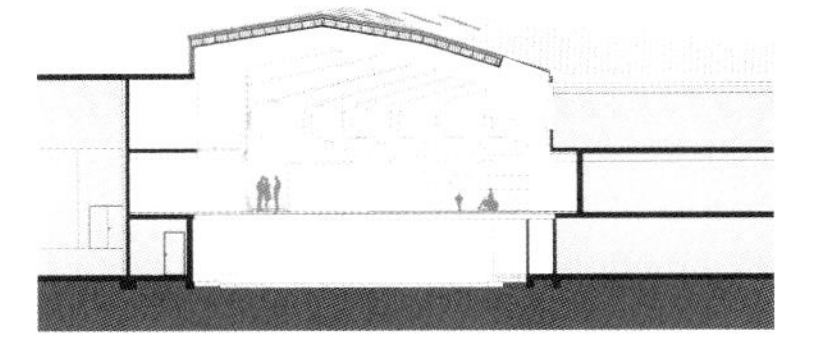

Cross section 2 横断剖面2

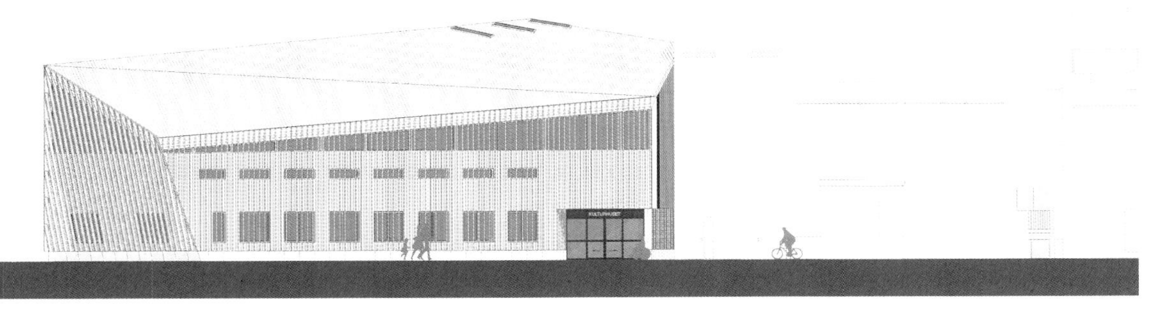

West elevation 西立面

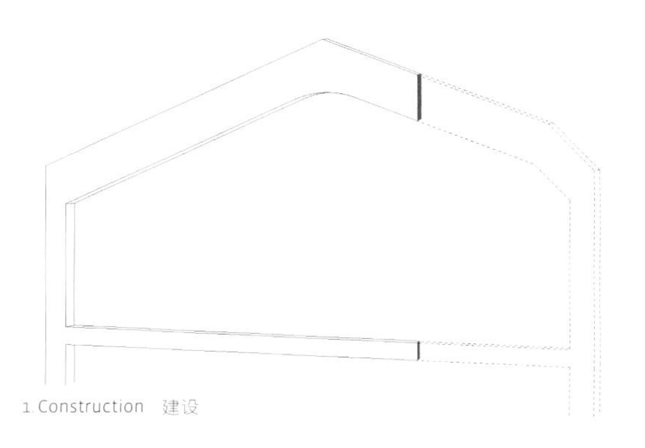
1. Construction 建设

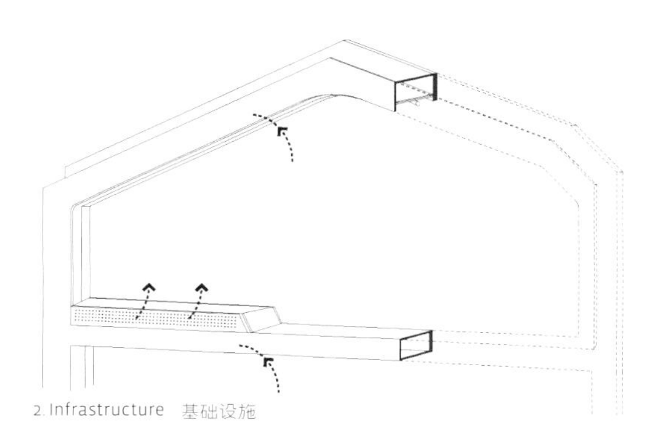
2. Infrastructure 基础设施

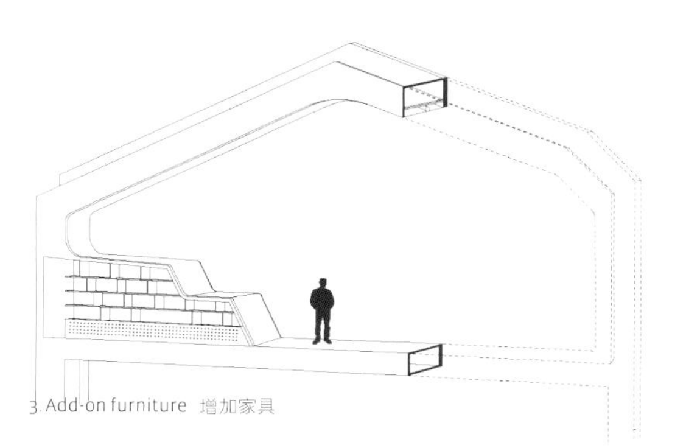
3. Add-on furniture 增加家具

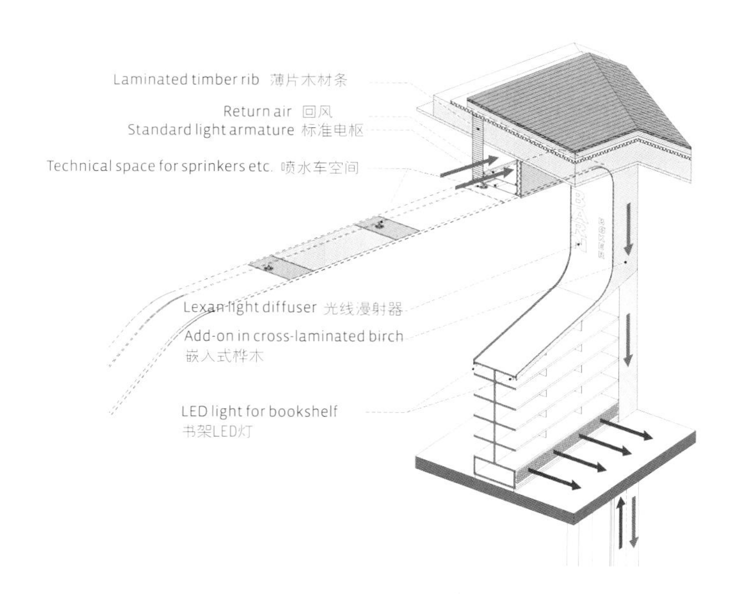
Laminated timber rib 薄片木材条
Return air 回风
Standard light armature 标准电枢
Technical space for sprinkers etc. 喷水车空间
Lexan light diffuser 光线漫射器
Add-on in cross-laminated birch 嵌入式桦木
LED light for bookshelf 书架LED灯

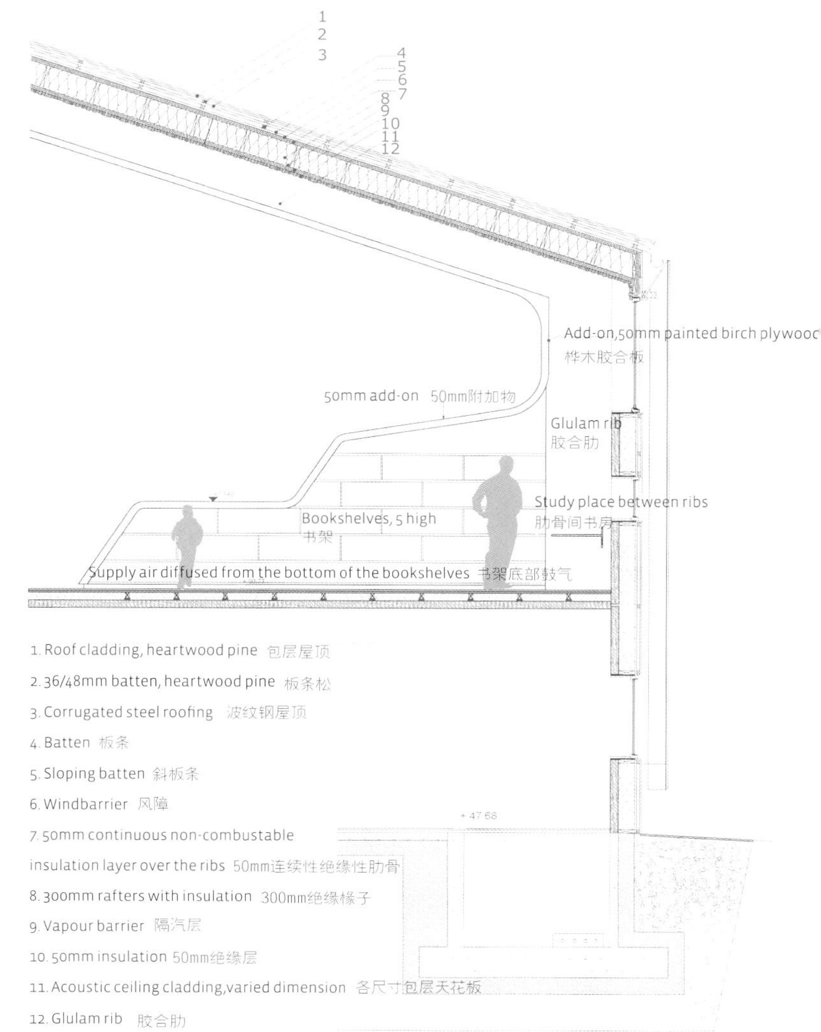
Add-on,50mm painted birch plywood 桦木胶合板
50mm add-on 50mm附加物
Glulam rib 胶合肋
Study place between ribs 肋骨间书房
Bookshelves, 5 high 书架
Supply air diffused from the bottom of the bookshelves 书架底部鼓气
1. Roof cladding, heartwood pine 包层屋顶
2. 36/48mm batten, heartwood pine 板条松
3. Corrugated steel roofing 波纹钢屋顶
4. Batten 板条
5. Sloping batten 斜板条
6. Windbarrier 风障
7. 50mm continuous non-combustable insulation layer over the ribs 50mm连续性绝缘性肋骨
8. 300mm rafters with insulation 300mm绝缘椽子
9. Vapour barrier 隔汽层
10. 50mm insulation 50mm绝缘层
11. Acoustic ceiling cladding,varied dimension 各尺寸包层天花板
12. Glulam rib 胶合肋

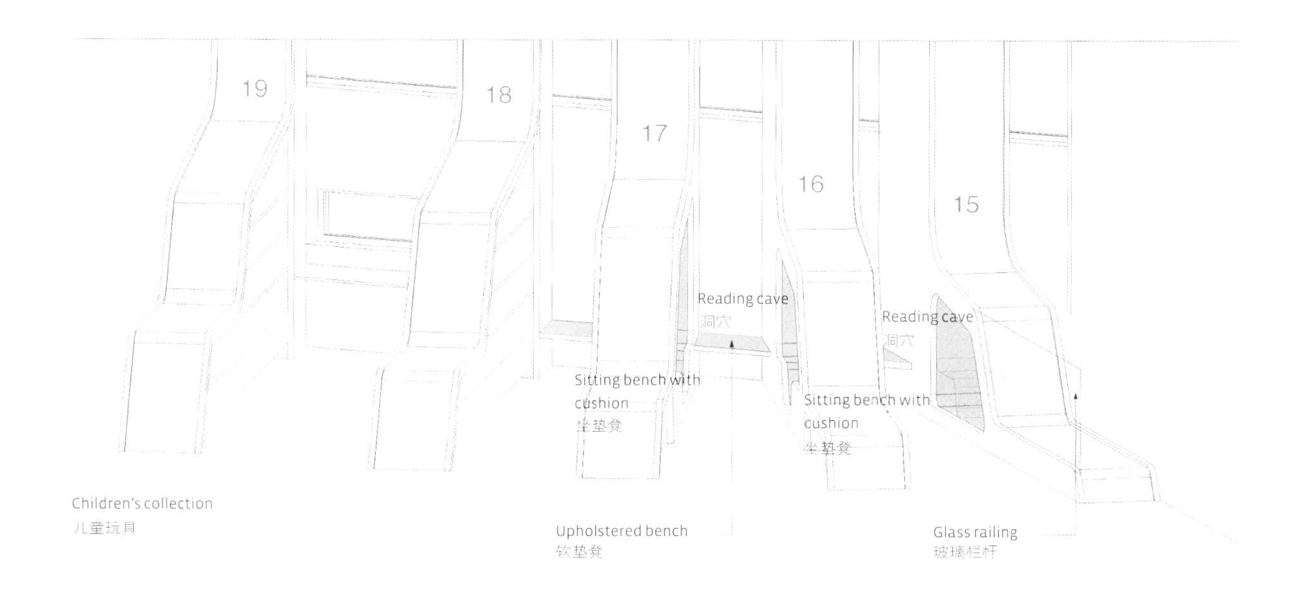

这座三角形的图书馆在城市中显得格外亲切，出入的便利以及靠近街道处低矮的建筑高度让人觉得格外舒适。整个建筑内部共有27根"肋"，这些"肋"传达出屋顶的几何形状，起伏的阵列构建出开放空间，也由此使建筑顶部起着转折的自然过渡，一方面考虑到相邻建筑，另一方面为了更好地融入自然环境。在另一侧，建筑外墙平面装有垂直遮阳板，为内部空间提供柔和的自然采光。

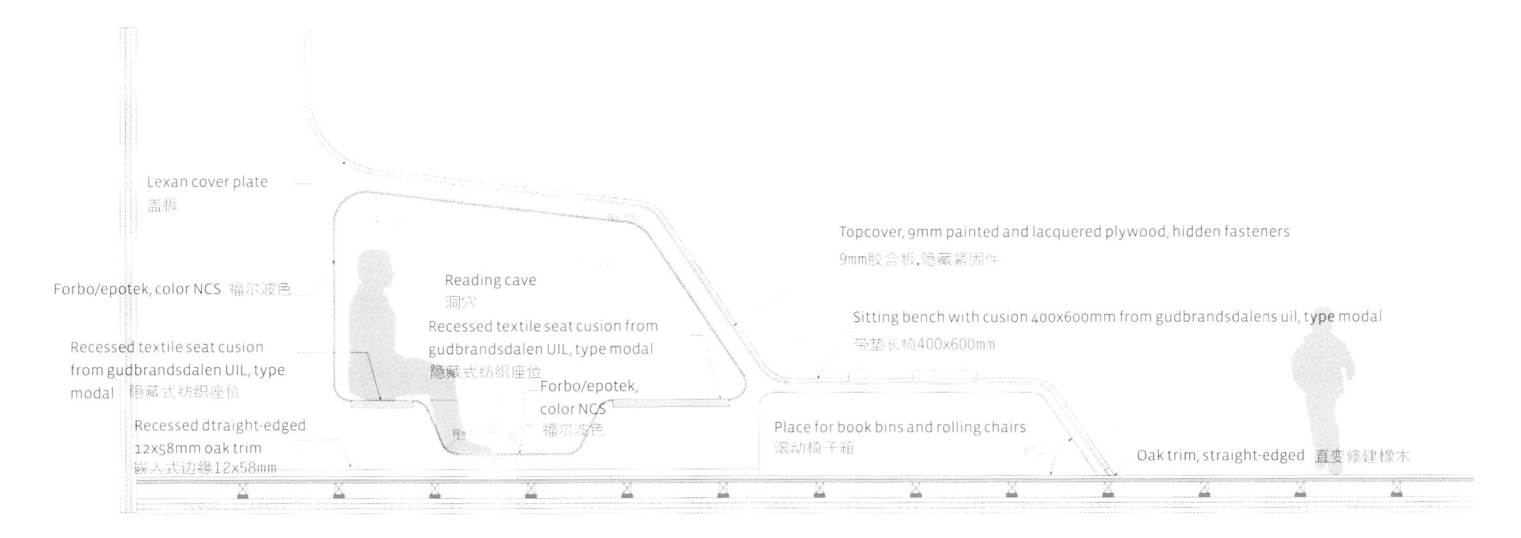

Detail section through reading cave 剖面细部

Within the center, there are four buildings of unique shape, composed of several triangle planes. Its special image appears in a totally new way, namely with the landscape surrounding architecture.

场地中心布置的4栋异形建筑，表面由多个三角形构成，其特殊的形象让建筑以新颖的面貌示人，更将建筑景观化，形成低密度感。

In the project, the façade is purely made of several triangle planes among which, spaces are distributed and form the correlative relationship with the buildings through the upper slanting volume. The underground entrance is a transparent glass box with a light box above, meeting the requirement of energy-saving. The architectural surface structuralizes the volume and extends into the glass lighting box, creating an effect of combining architectural volume, material and space.

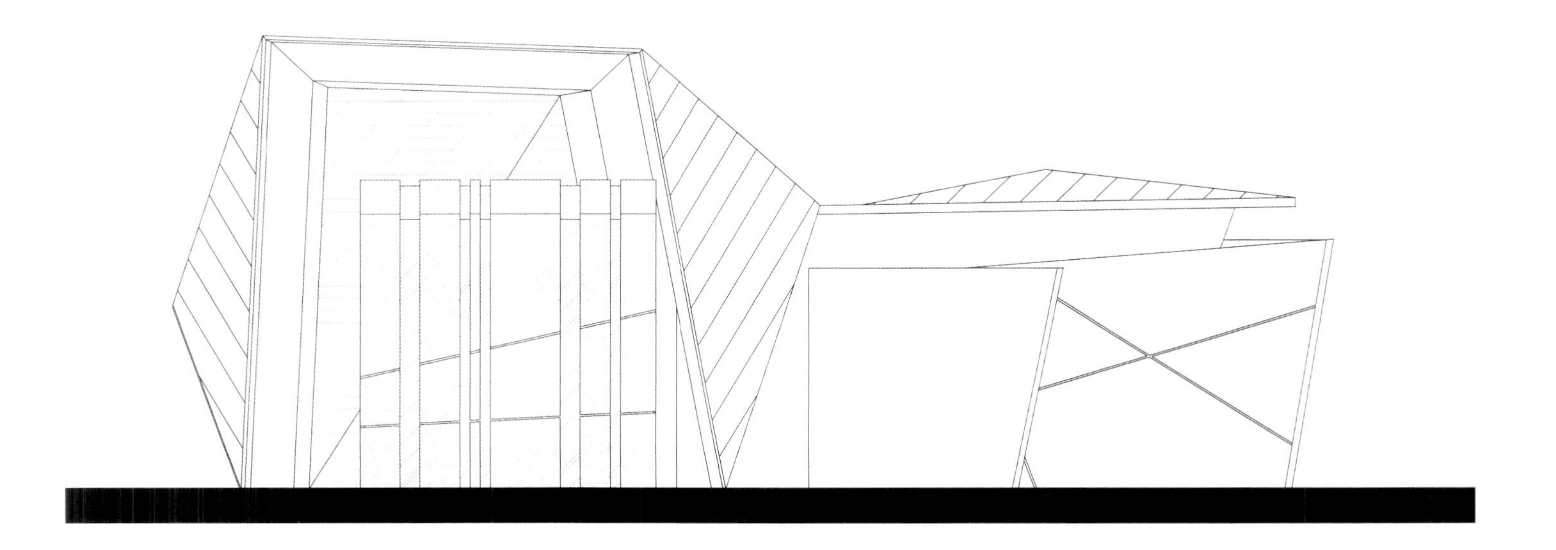

本案建筑表面为纯粹的三角形堆砌的平面形塑，并将空间内容分置其中，再藉由上部量体的倾斜错置构成建筑与内部空间的相互关系。地面层入口量体为一个通透的玻璃光盒，而三角形平面置于光盒之上，形成的“盒中盒”的概念达到了防止西晒的节能设计要求。建筑表皮上将量体片状结构化，并延伸进下方的玻璃光盒之中，能营造出串联建筑量体、材质与空间的效果。

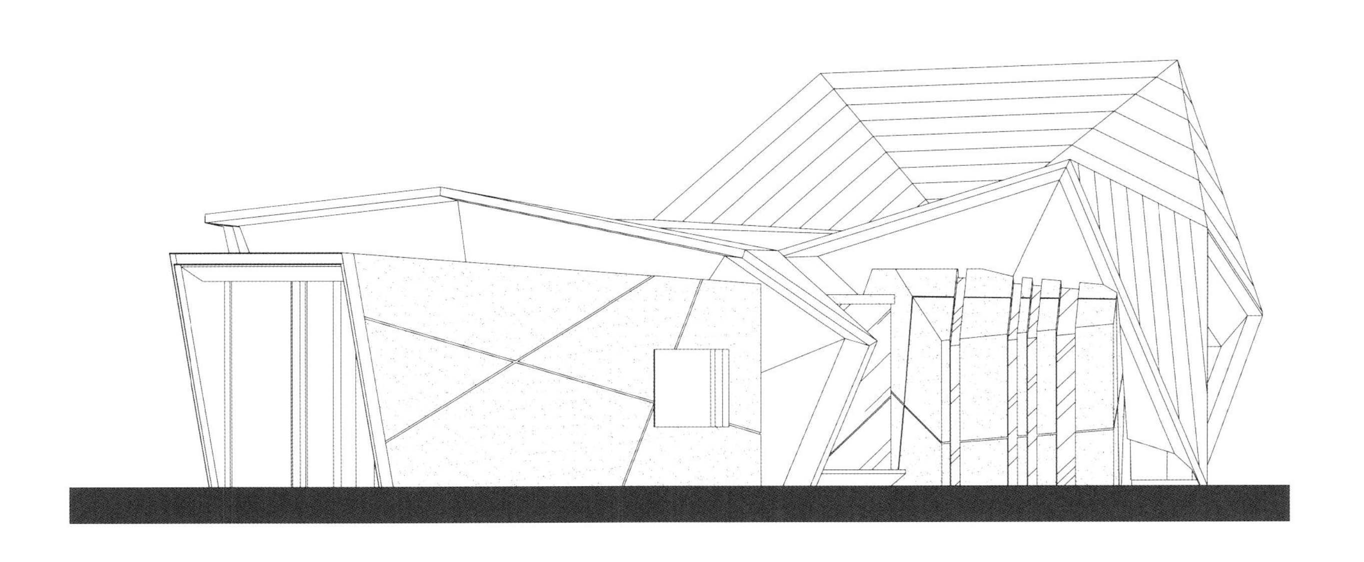

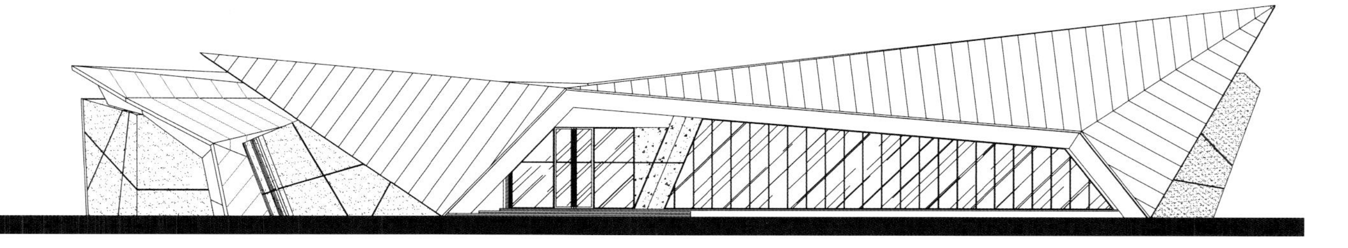

60-77-120坪 666-1058

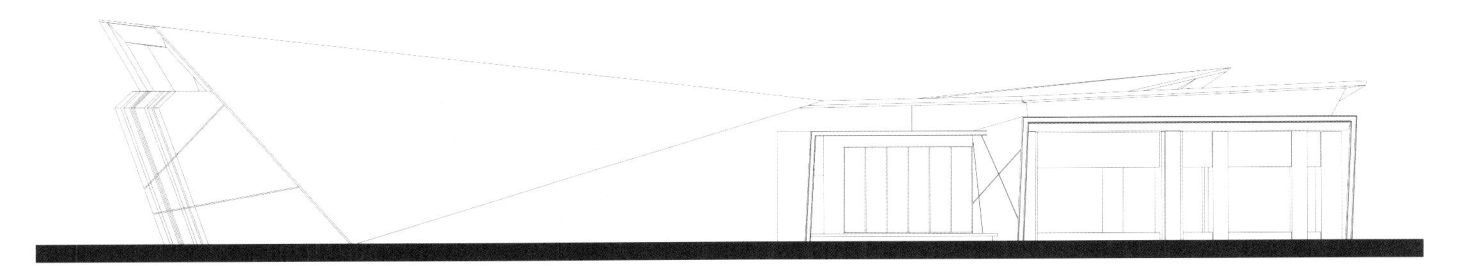

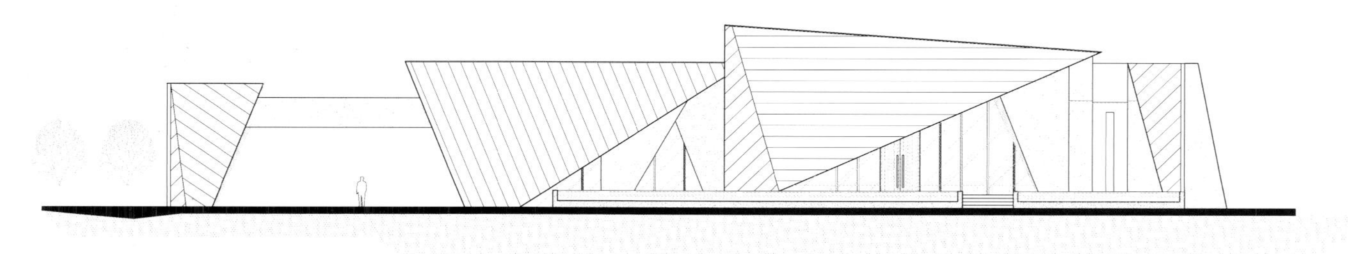

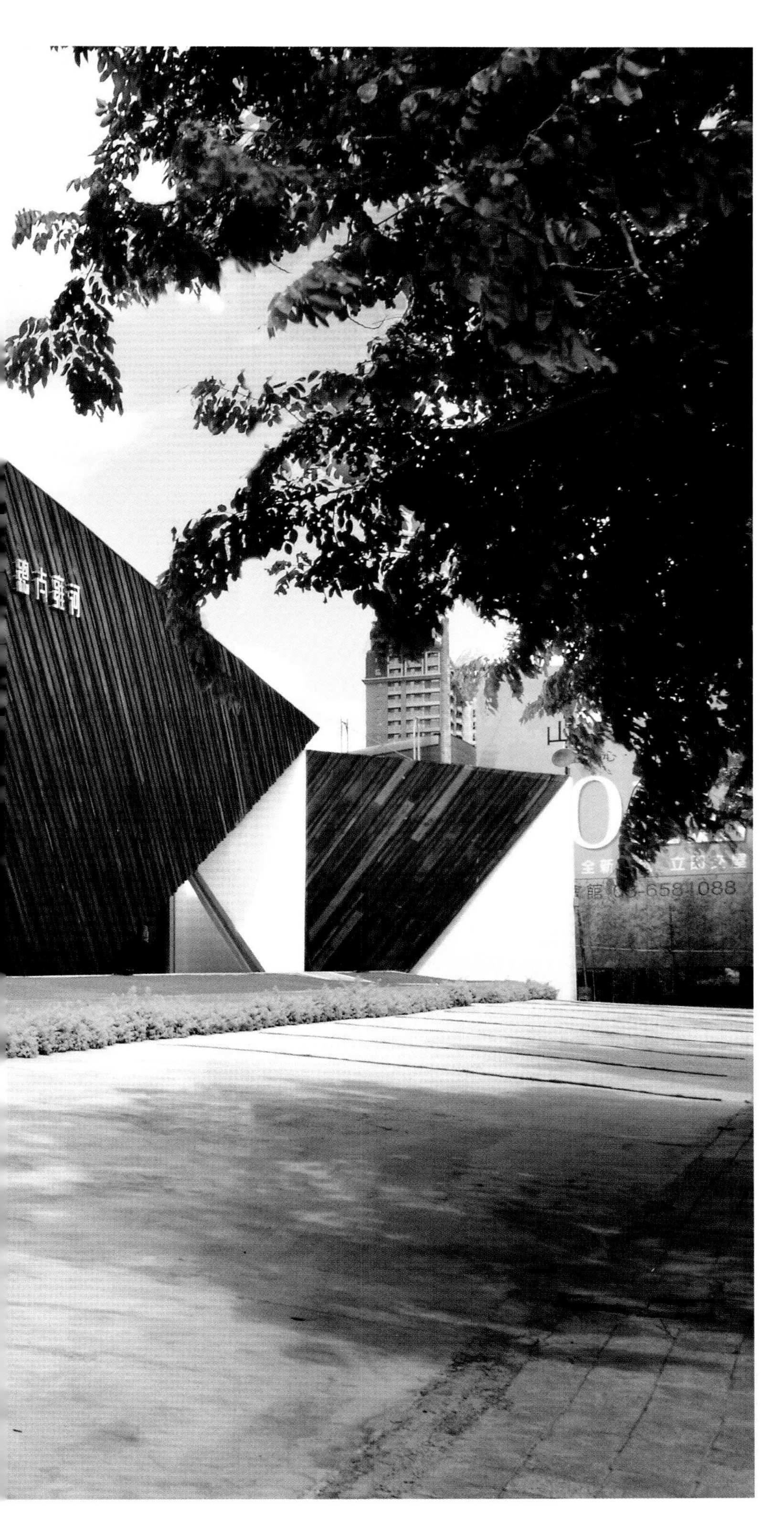

Triangles in different placements may give people different induction. Although triangles are stable, they give a stable feeling when the top is above while the top is at the bottom, the triangles will be tensioned as the trees and the ground. Thus, the architecture of differently placed triangles will form the interesting façade, giving a striking visual experience.

正反三角形，摆法不同，带给人们不同的心理诱导。尽管其本身是稳固的，顶点在上面时，由于考虑以广阔的接触面向大地传递重力，所以会产生稳定感；顶点在下面时，重力集中于尖端，与树木一样和大地构成一体，会产生强烈的紧张感。正反三角形组合而成的建筑造型，则会形成有趣的立面表达，给人两种对比鲜明的视觉体验。

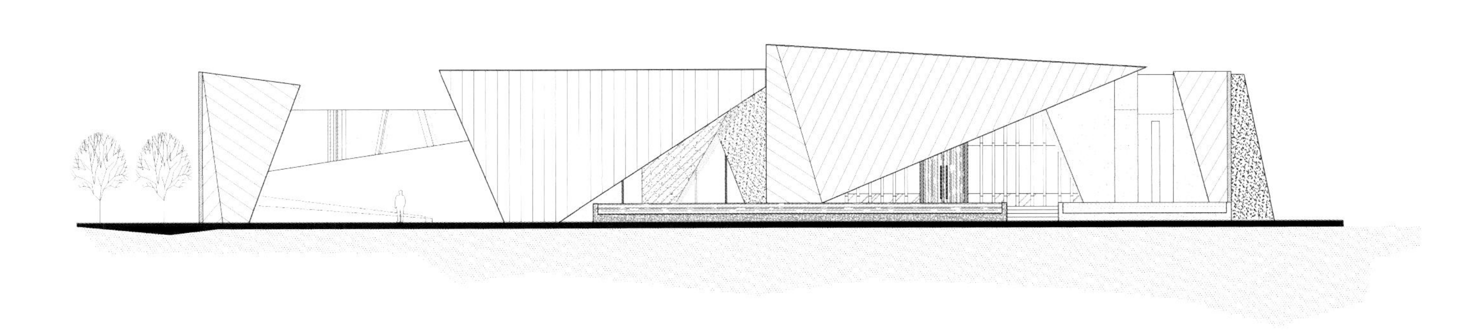

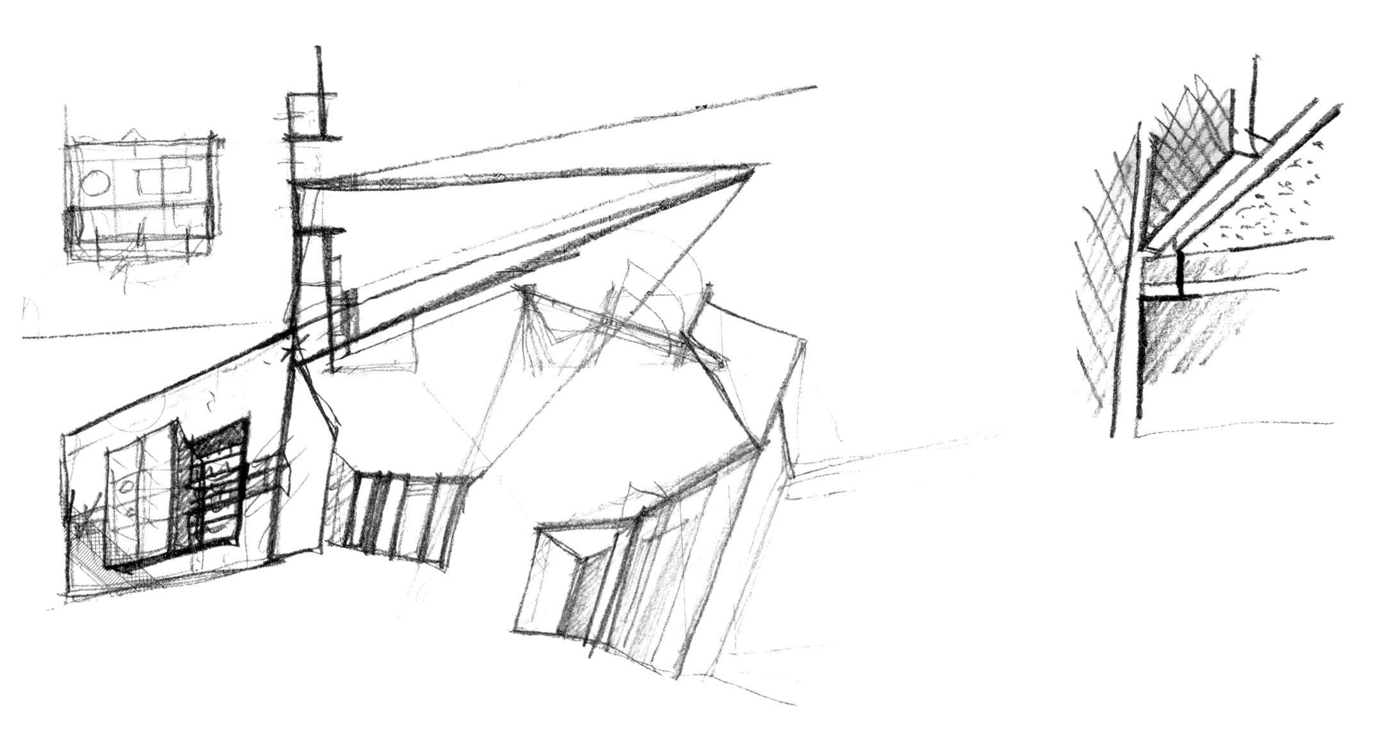

The buildings mainly with the triangle façade mostly bear the practical functions, like load-bearing or lighting but its origin of ideology shouldn't be neglected. In fact, the triangle building with strong aesthetic is an eye-catcher. The building not only attracts people's attention but makes the building interesting. Its exterior is powerful in direction, with diverse, random and flexible sides, giving a stable sense.

以三角形态为主要立面造型的建筑，大多承载着实在的功能，如承重或采光，然而不应忽略它的意识形态的来源。事实上这种带有强烈美学特征的三角形建筑空间是有着某种巨大的魔力以及某种莫名其妙的吸引力量。它既能够吸引人们的注视，又能使建筑显得更加的新颖而有趣。其外部具有强烈的方向性，边角表现出更多的多样性、随意性和灵活性，给人稳定、不易变形的稳固感。

JANSEN

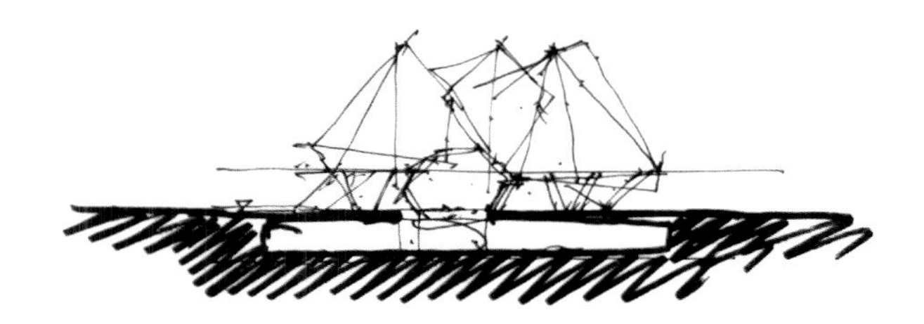

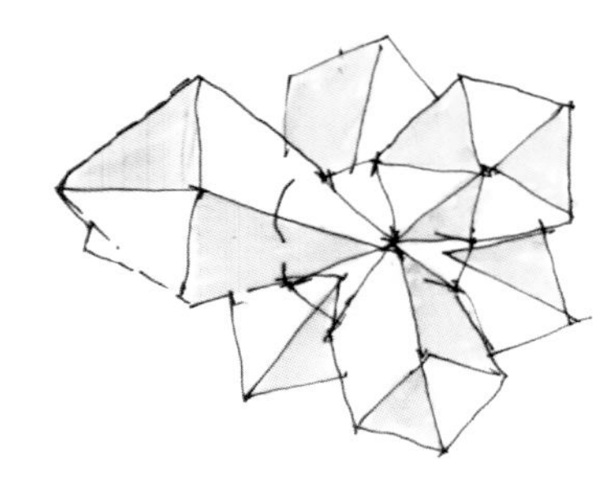

The triangle architectural shape is representative and inspirational. Its striking feature takes on vigorous strength. Here the regularly placed triangle symbols give rhythm to the architectural façade. Similar triangles form the accumulated shape, diversifying the architectural outline.

三角形的建筑造型，能够具有某种象征和启发的力量，其外显的特性使得以尖角符号为特征的立面呈现出勃发向上的气势。在这里，按照一定规律排列的三角形符号，使得建筑立面产生了韵律与节奏感。相似的三角形依次组合形成积聚的形态，让整个建筑造型更富有变化。

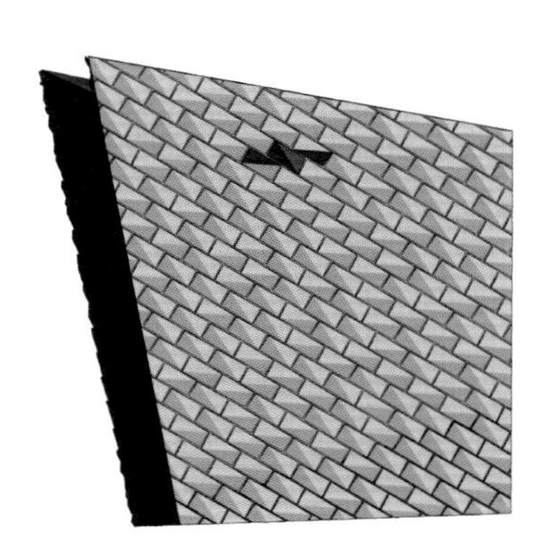

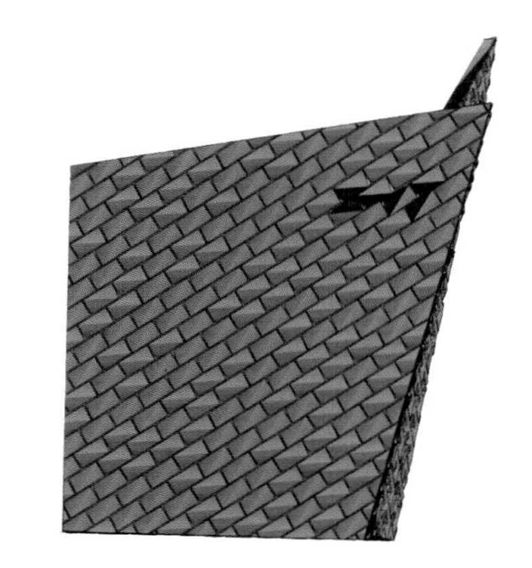

Vertical, or close to vertical panel arrangements are more rational but tend to be more static.
靠近垂直平面布置，布局合理且静态

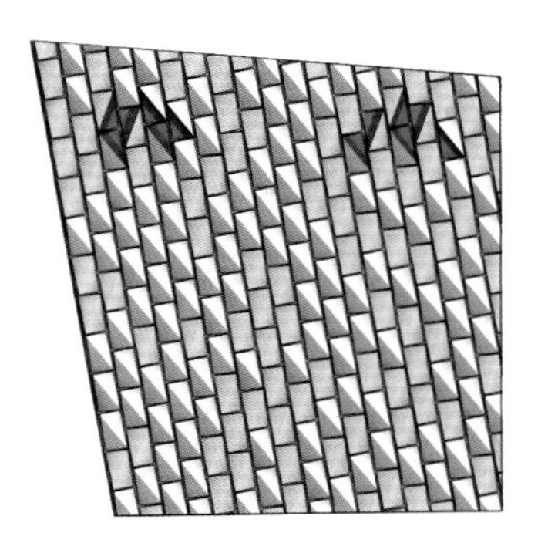

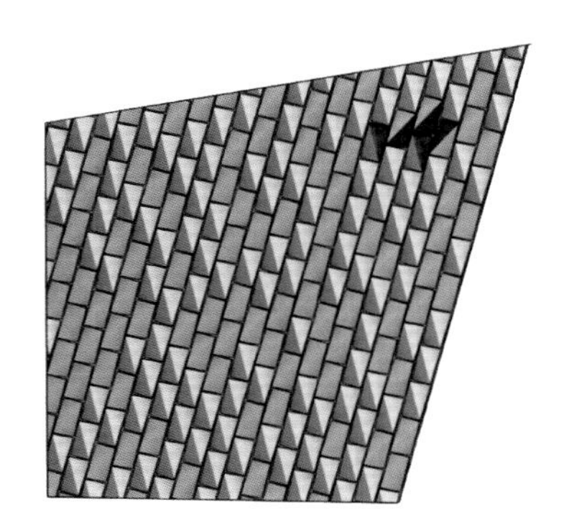

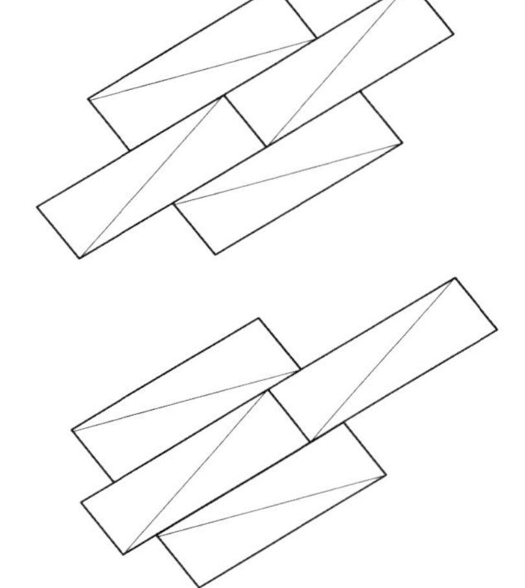

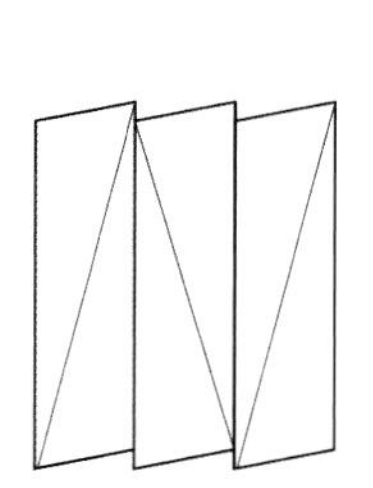

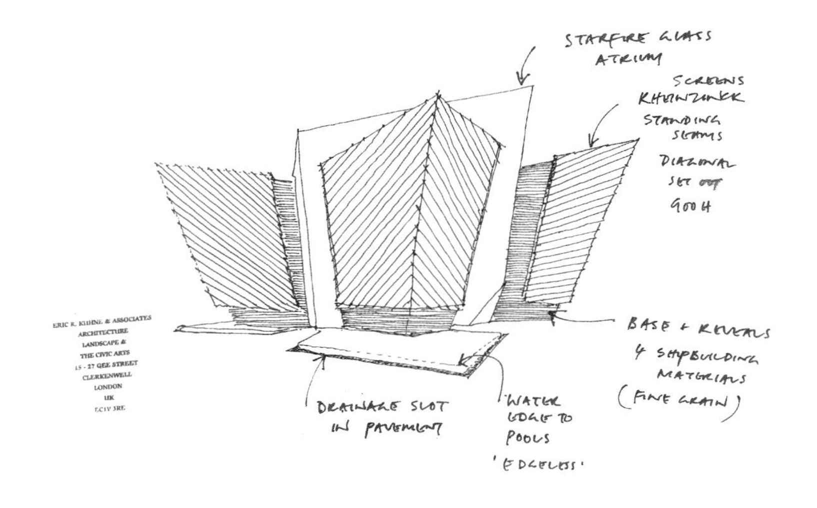

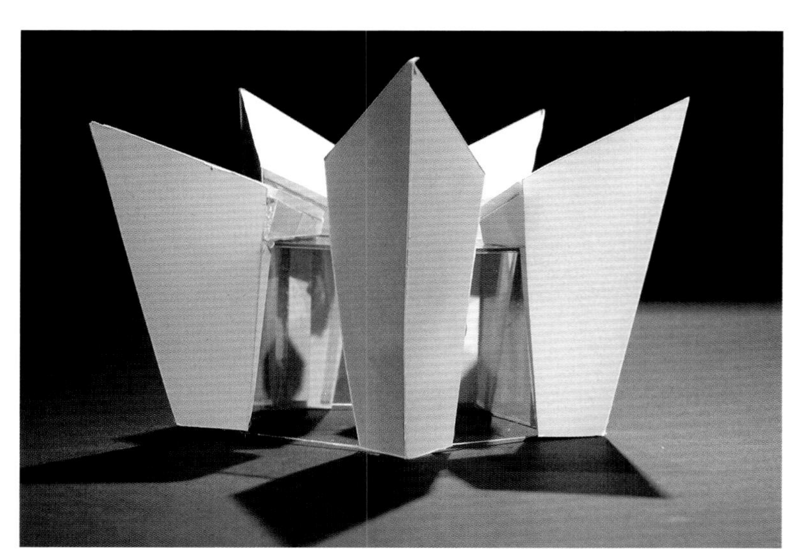

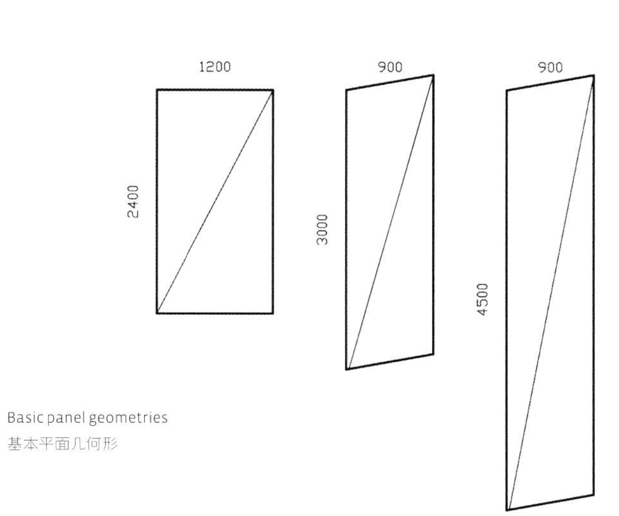

Basic panel geometries
基本平面几何形

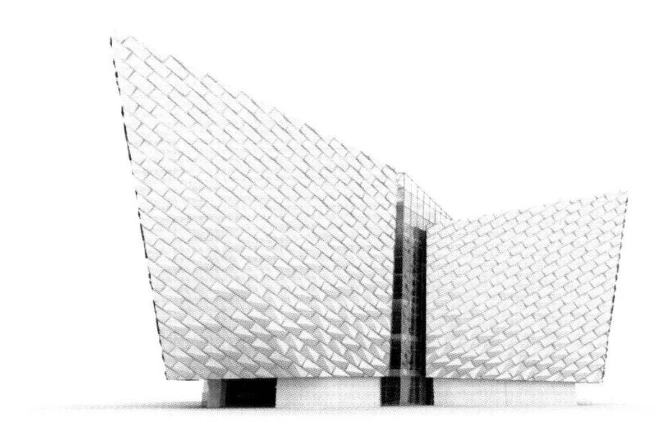

True elevations will never be seen; especially from within memorial place.
从纪念馆位置无法呈现立面实图

TITAN

WINGET

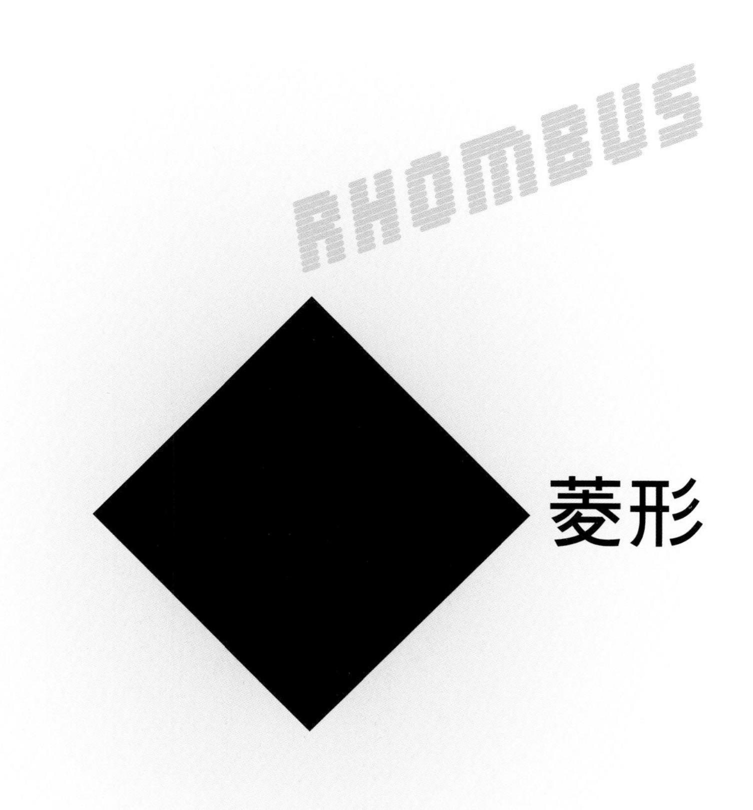

菱形

POWERFUL, POINTED, RESERVED, WITH UNIQUE GEOMETRY BEAUTY

硬朗锋利、含蓄内敛，流露独特的几何美感

A rhombus is another way of placing squares and its four angles represent the inward balanced power. Therefore its balance is determined by the four directions of up and down, right and left. The prolonged lines are suit for the skeleton of continual patterns which breaks through the dullness of grids.

A rhombus is often associated with a diamond whose design is the flexible application of the geometry. But it's hard to relate a rhombus with architecture and to imagine what miracles will be worked out. Its geometry contains rich aesthetic knowledge which is favored by designers of different schools. From the architectural outline to the details, outside decorations, rhombuses are seen almost everywhere. When used in the façade, a rhombus varies like a relief, rich, yet reserved with unique beauty.

In modern architecture, rhombuses are increasingly employed in the façade. Its regular yet unique geometry gives a symmetrical, harmonious beauty. Its pointed angles represent enterprising spirit and hope, just like the Glass Pyramid of Louvre, the conical American Building. Their sharp facades takes on hope of breaking through everything and vivid beauty, with powerful visual shock. In designers' eyes, a rhombus is usually a difficulty due to its rigid lines and acute angles that represent powerful tension and wildness that designers often find difficult to dominate. Therefore, the successful rhombus buildings display the designer's exceptional conception abilities.

菱形，是方形的另一种摆放方式，由于视知觉根据力的作用来确定上下方的位置，而菱形四个角体现出由中心向上下左右的均衡的力，所以它的平衡感是由上下左右四个方向来确定的，它的四条边的延长线，极适合做连续纹样的基本骨格，并且其形成的网格纹突破了井字格的单调与平均。

提起菱形，人们总是会情不自禁地联想到钻石，钻石设计便是对菱形的最巧妙的应用。可是人们很难想象将菱形应用到建筑设计中，将会创造怎样的一种奇迹。几何形态包含着丰富的内涵与审美情趣，因而常常成为不同流派设计师所青睐的造型语言。大到建筑整体造型，小到各种细部以及外立面装饰元素，菱形几何形态似乎无处不在。菱形几何图形运用到建筑外立面中，其凹凸变幻如同浮雕，组合起来变化丰富却又含蓄内敛，流露出独特的美感。

在现代建筑设计中，菱形被越来越多的设计师所发现，并应用到外立面设计中，其规则又显独特的平面形态，往往能带给人一种对称、和谐之美，而其尖锐的角，又象征着一种精神与希望。例如法国卢浮宫广场上的玻璃金字塔、美国旧金山的锥形建筑——泛美大厦，尖锐的立面形象，给人一种冲破一切的希望，有种淋漓尽致的美感，具有强烈的视觉冲击。但菱形往往又是设计师眼中的一个难题，因为菱形硬朗的线条、锋利的锐角，往往体现了强大的张力与野性，很难被一般的设计所驾驭。然而也正因如此，那些成功的菱形建筑作品恰恰体现了设计师与众不同的设计构思与强大的抽象思维能力。

Heaven 上天
Sky 天空
Earth 地球

Sky city 空中城市
Leos of education 里欧教育

The solid rectangle form contrasts with the graceful rhombus outline. In the daytime, the rectangle volume stays peacefully in the city's hug, quiet and firm. In the night, it becomes a lighting object when illuminated. The pure and simple geometry brings its uniqueness into the fullest play. The minimum principle is a trend in architecture, as can be seen in the flexible use of environmental elements, display of steel and shading, pure pursuit of basic geometry forms and rectangles.

结实而厚重的长方形造型与流畅而轻盈的菱形外框架形成了鲜明的对比。白天，它静谧地躺在城市的怀抱中，淡定而坚实；到了夜晚，在灯光的照射下，它就成了发光体。纯粹而简洁的几何形在灯光映照下将其特有的形式表现力发挥到了极致。这种对现场环境元素的巧妙转化与运用，对钢材及光影的塑造和表现，对基本几何形的纯粹性追求以及对矩形结构的精致表现中，都可以看出建筑设计对极少主义的倾向。

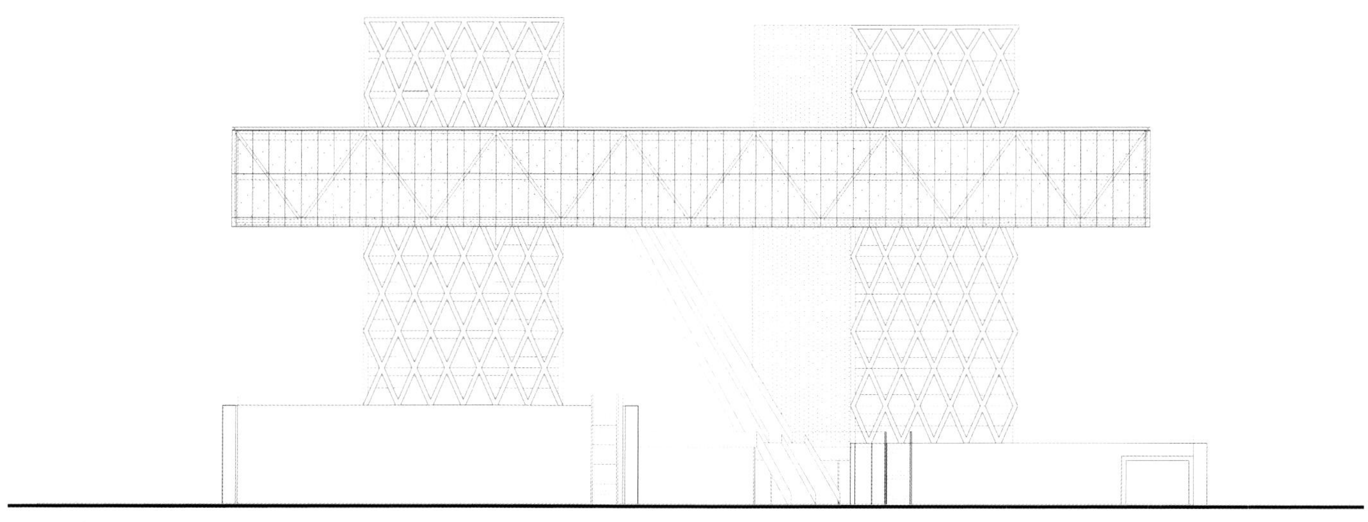

West Elevation
西立面

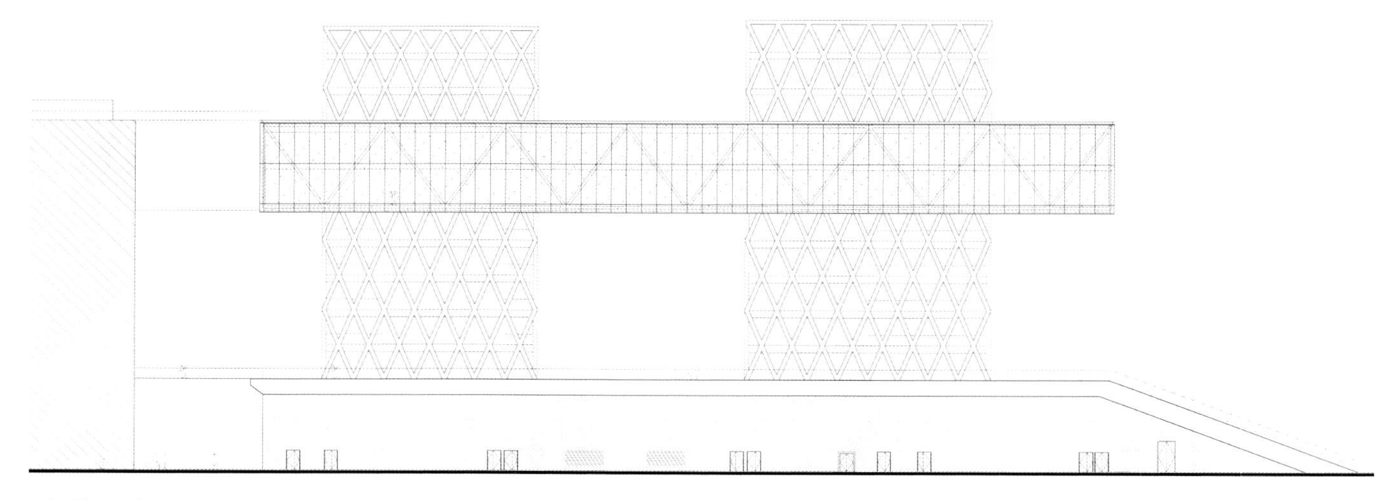

North Elevation
北立面

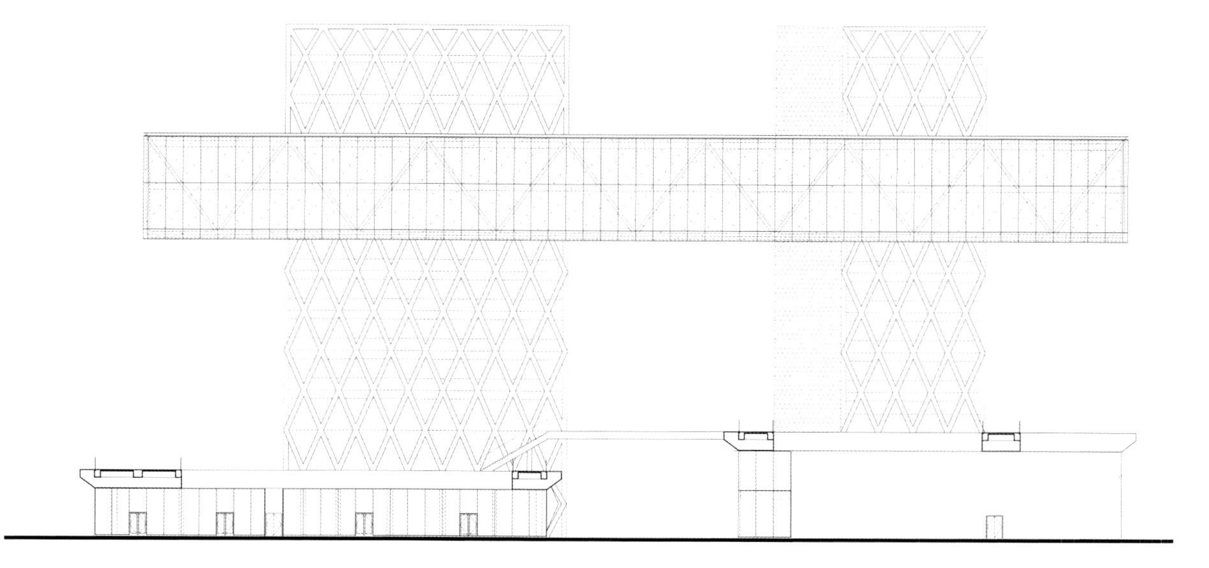

East Elevation 东立面

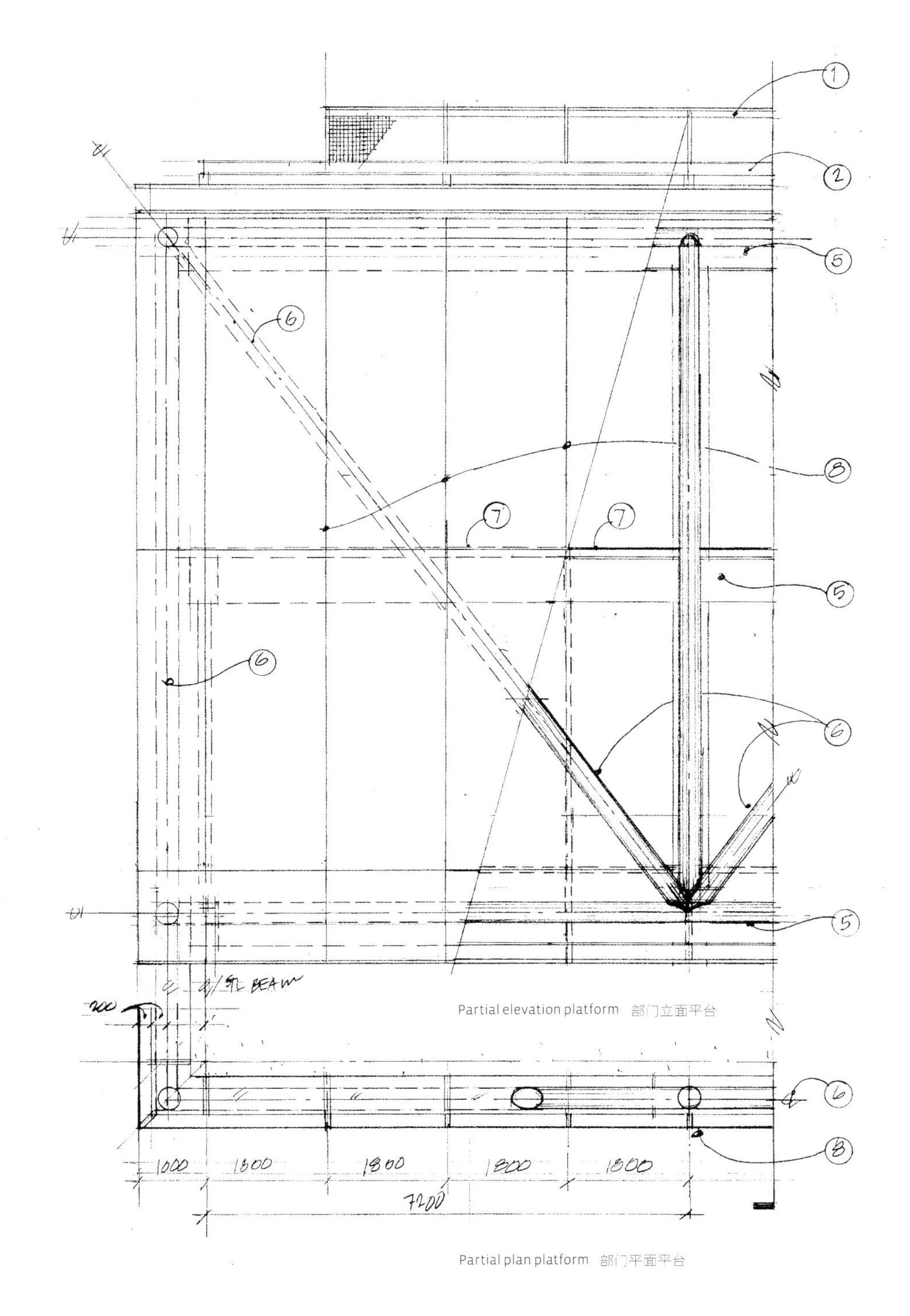

Partial elevation platform 部门立面平台

Partial plan platform 部门平面平台

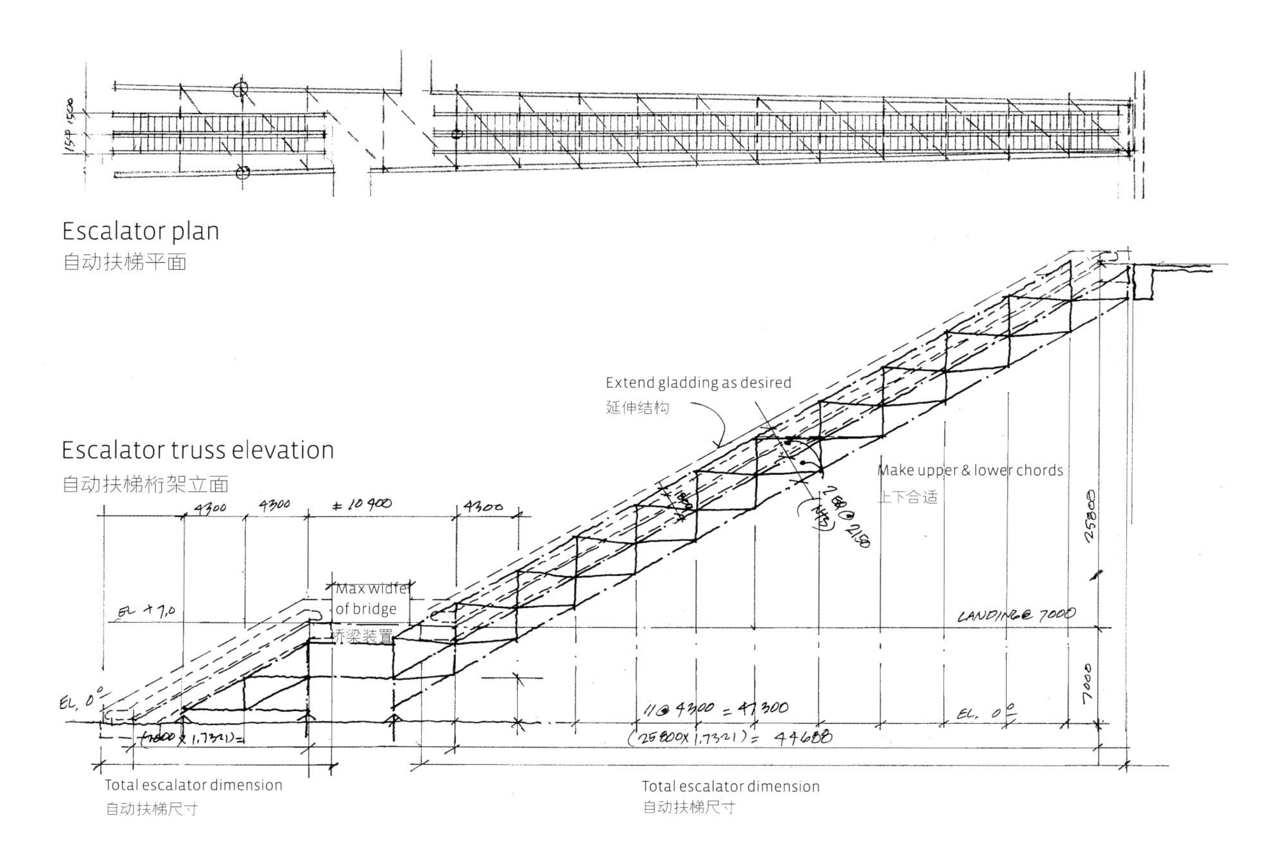

Escalator plan
自动扶梯平面

Escalator truss elevation
自动扶梯桁架立面

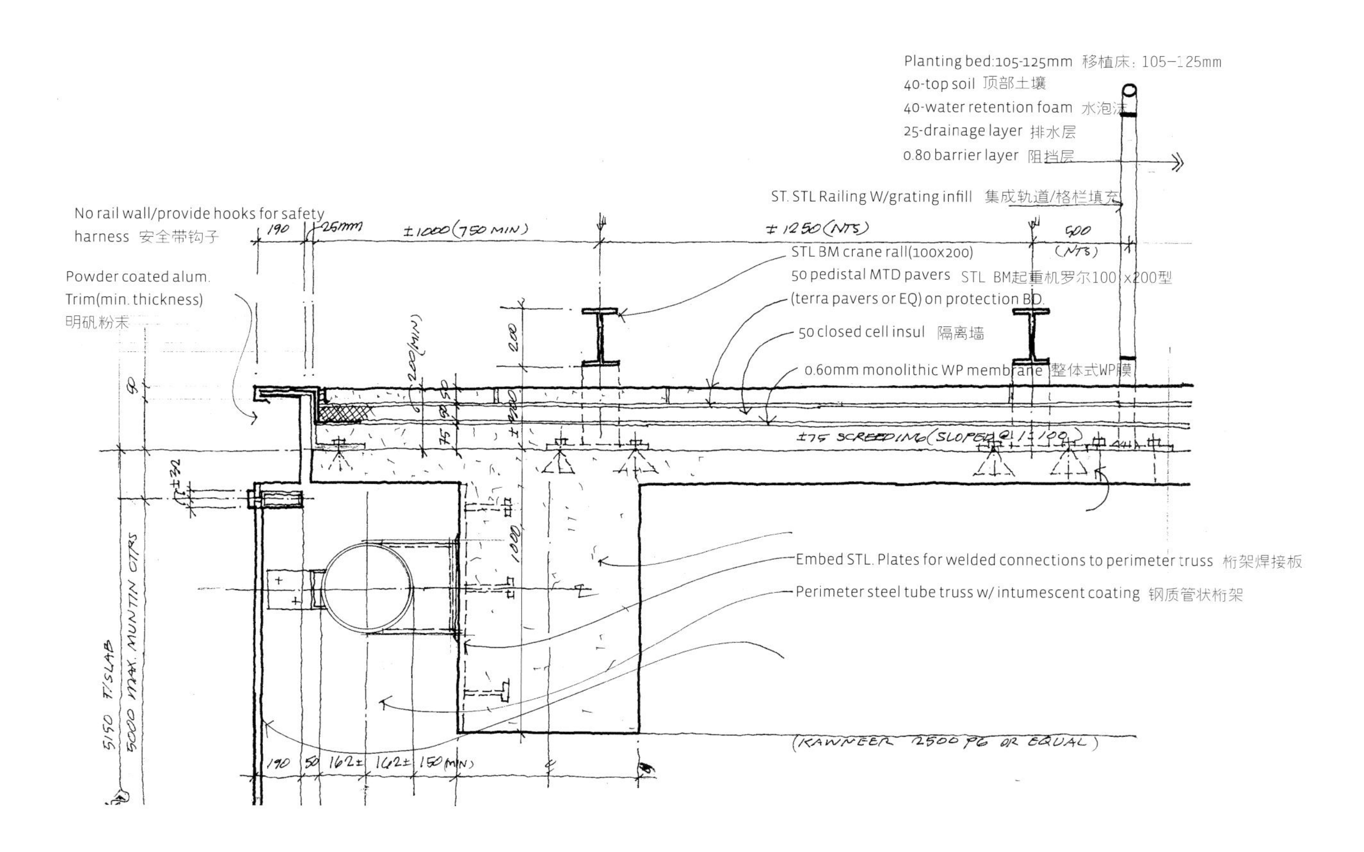

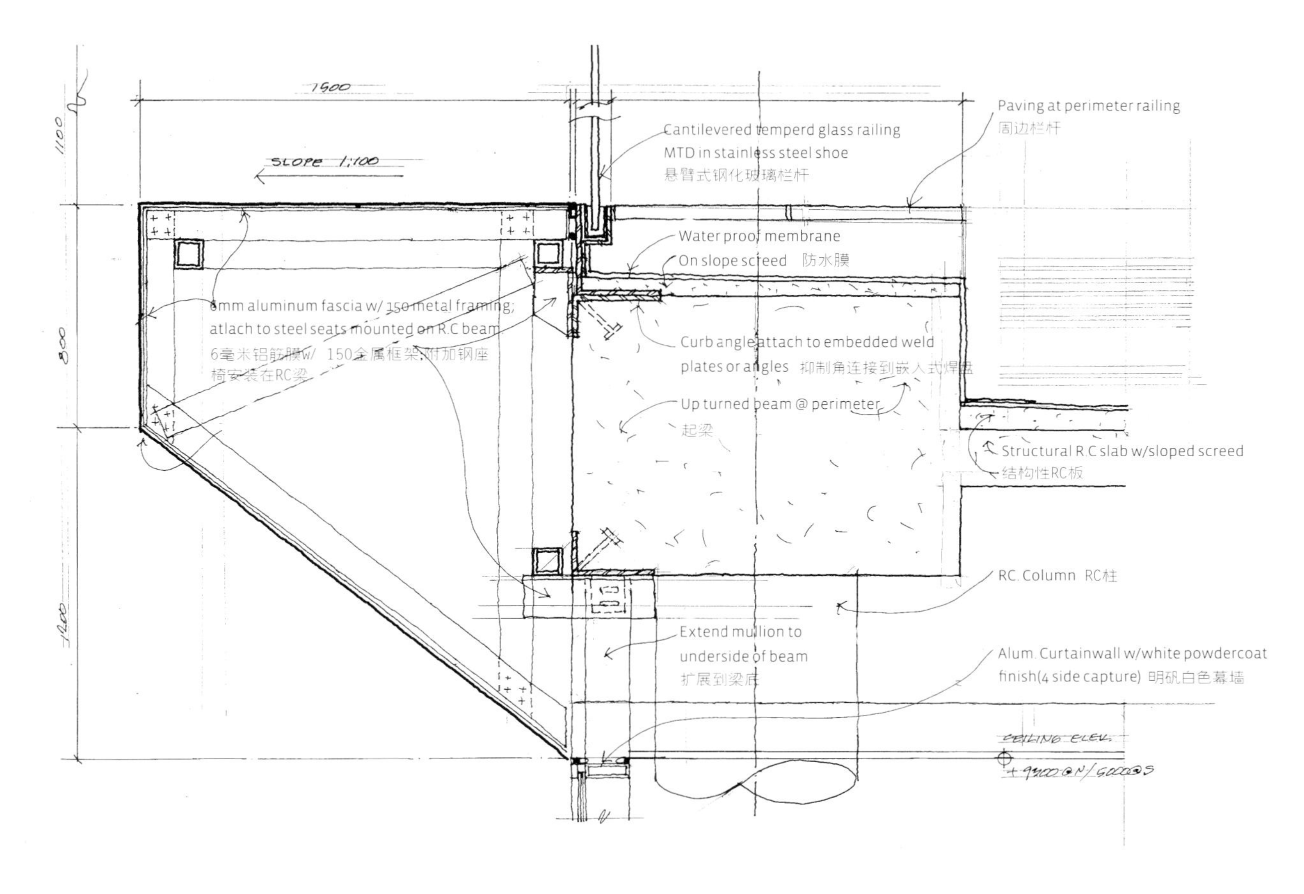

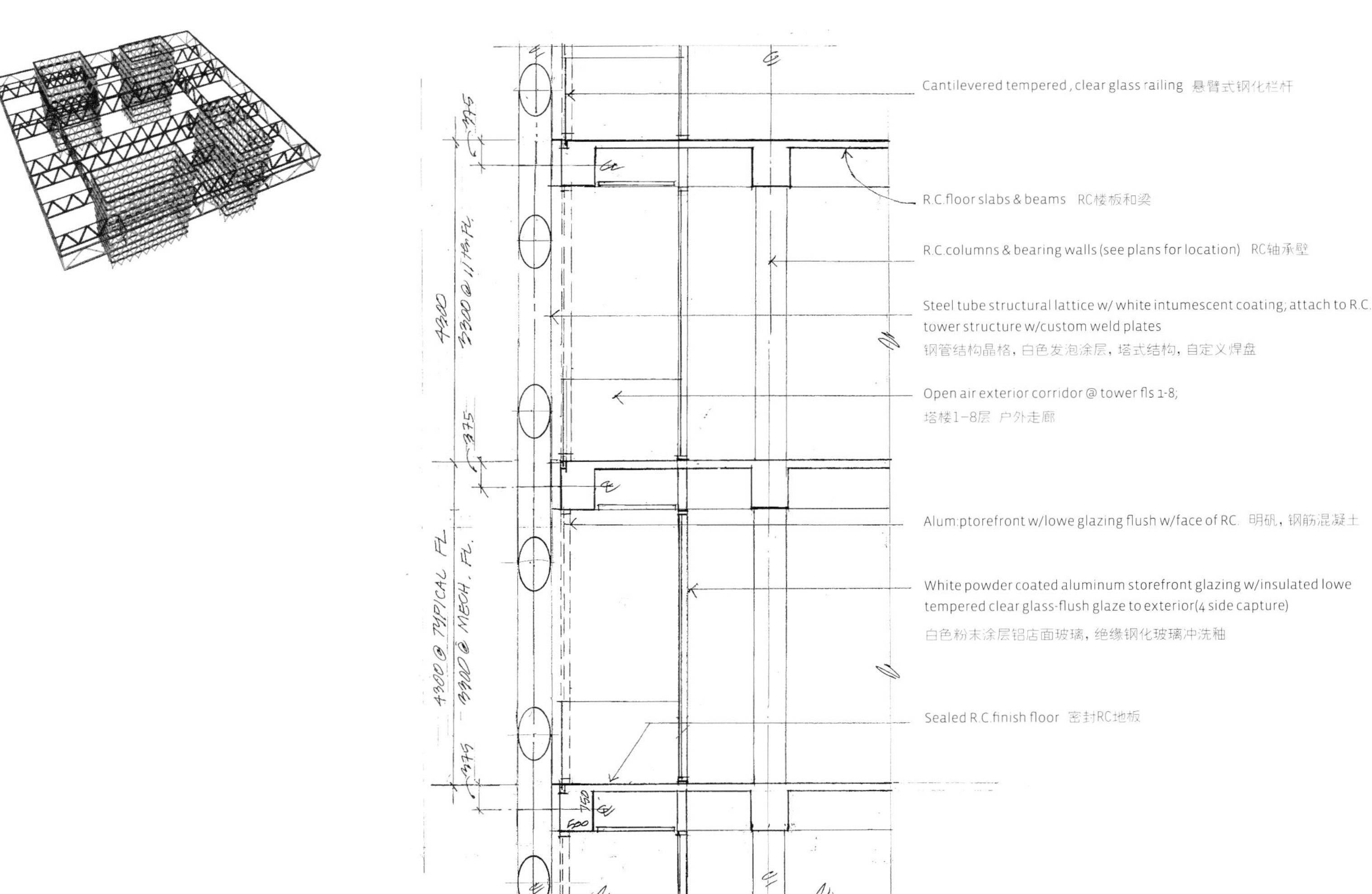

Partial section tower lattice 塔阁部分剖面

1 Stainless steel railing W/ crating infill 不锈钢轨，装箱加密

2 STL beam window wash crane rail 起重机轨道

3 Planting bed: 植物床
40-80 top soil 土壤
40-water retension foam(grodan or EQ) 水自留泡沫
25-drainage layer 排污层
0.80 barrier layer 阻挡层

4 Service walkway: 维修走道
50-pedital MTD pavers on protection BD 铺路材料
50-closed cell insulation 隔离层
0.40-monolithic WP, membrand 单片WP
75-sloped screeding 倾斜抹平

5 RC beam & FL& roof slabs 钢筋混凝土梁与屋顶板
W/ intumescent coated steel tubes 钢材覆盖管

6 Perimeter steel tube truss w/white intumescent coating
钢管桁架

7 1 HR horiz."fire glass" LAM 防火玻璃
Infill: flush w/ floor@ 填充物
Perimeter btwn slab & curtain wall muntin 幕墙窗隔条

8 Unitred aluminum curtainwall(white powder coat finish); 4 side silicone glalink.
铝幕墙

9 Soffit:powder coated aluminum panels w/concealed fasteners 铝面板
w/rigid suspension system 刚性悬挂系统

Partial Section @ Platform
平台部分剖面

WELCOME TO QUANG TRUNG SOFTWARE PARK

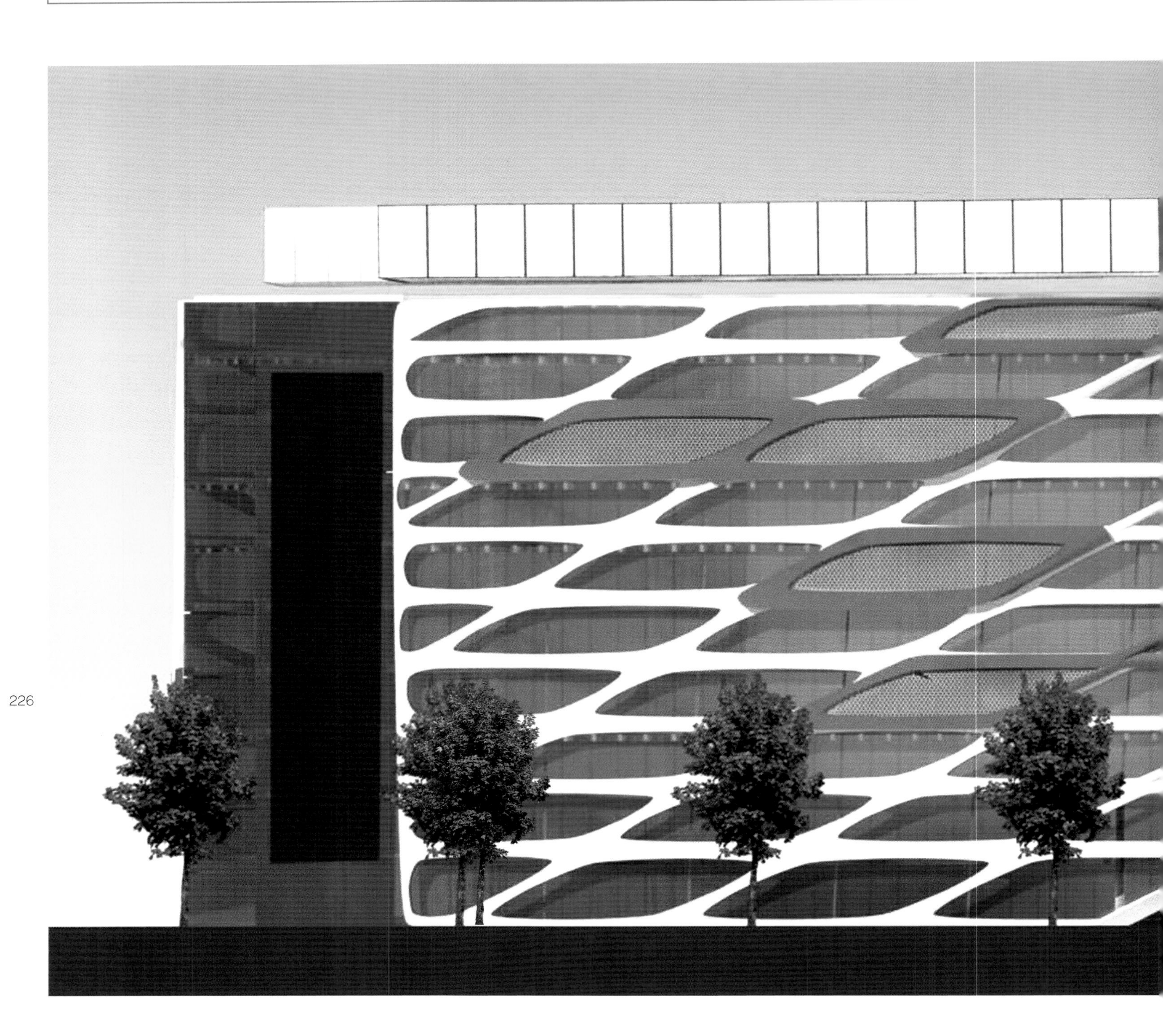

The architectural design is to maximally improve natural ventilation. The crossing axes on the surface form a unique façade. At night, with illumination, transparent and semi-transparent glass rhombuses form a dynamic surface.

该建筑被设计来最大限度地提高自然通风和空气流通，表面上纵横交错的轴线构成了独特的立面形象。透明与不透明的玻璃菱形平面的组合，到了夜晚，利用灯光的照射，形成动感的界面效果。

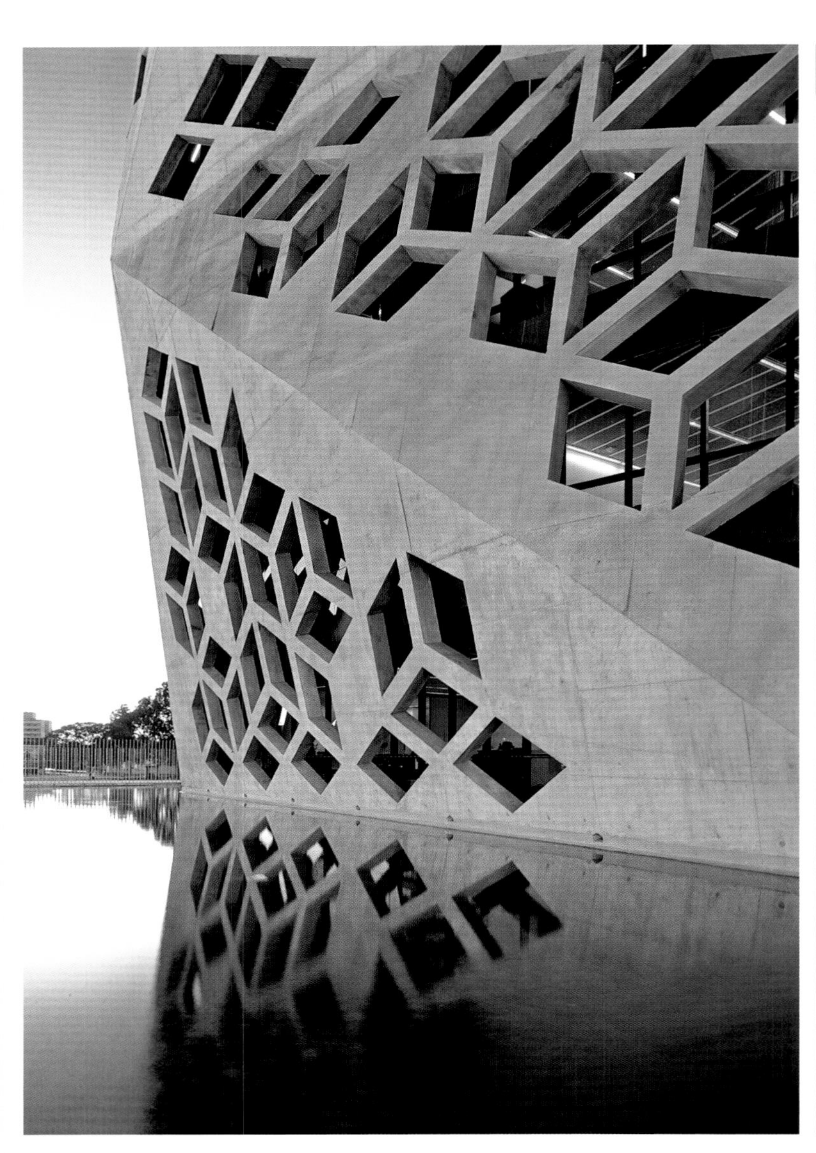

The building is a high-rise concrete structure. The 16-m building is folded in 20 degrees and the folding surface affects the top part. The triangular folds form special light shadowing to make the building more dimensional. The architect makes many diamond hollows to form the representative surfaces.

该建筑是一个中高层的混凝土棱体，在16米的高度上做了20度的折面，这个折面又影响到了上层形态。建筑的一系列三角形折面形成了特殊的光影，为了让建筑更具立体感，建筑师在三角形折面上做出了许多菱形的镂空，形成具有标示性的建筑表皮形象。

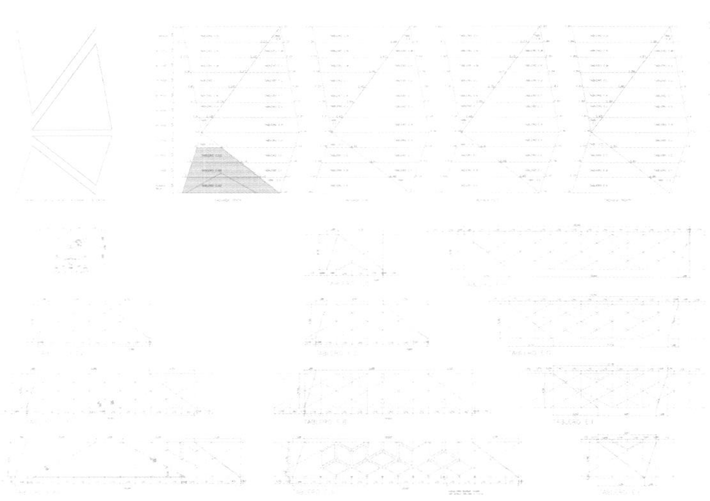

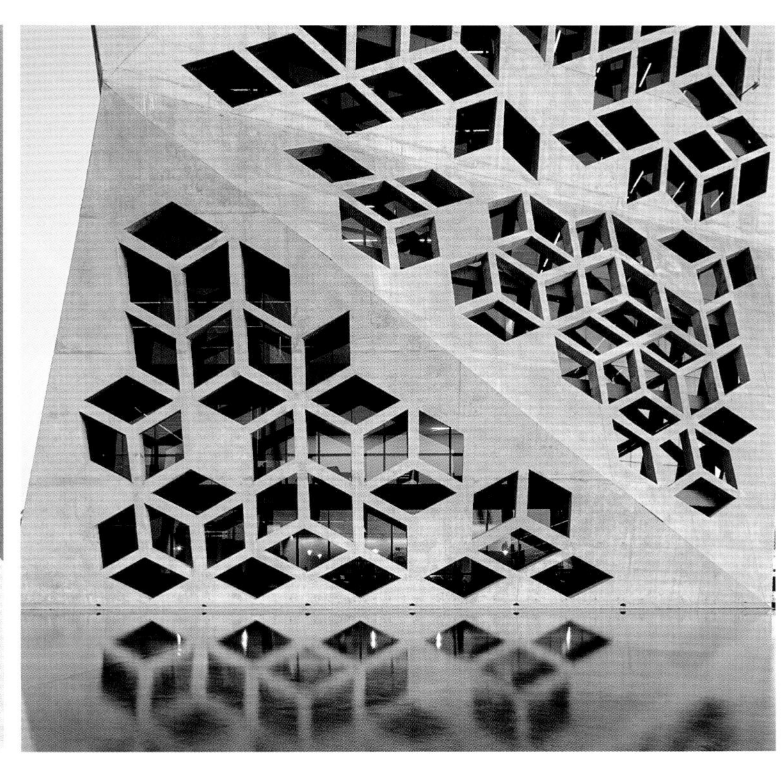

Fachade Este 东立面

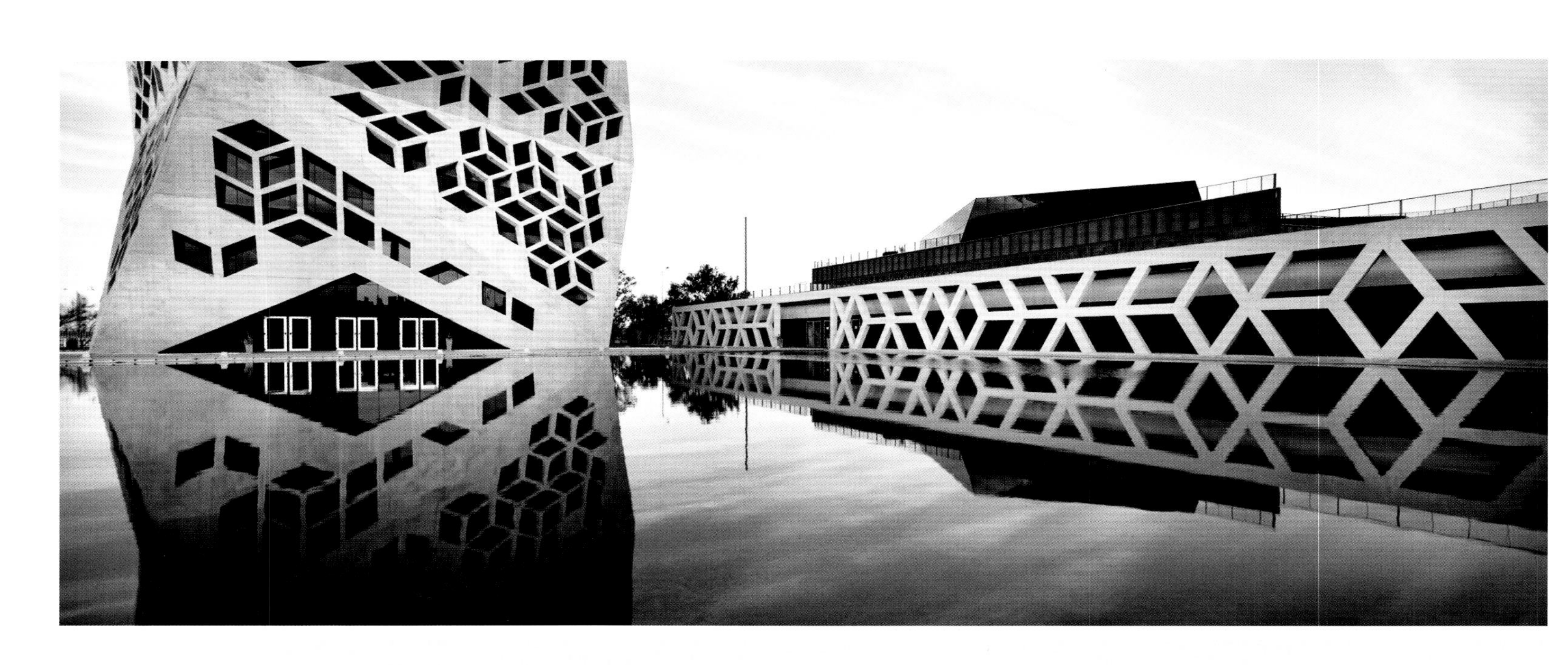

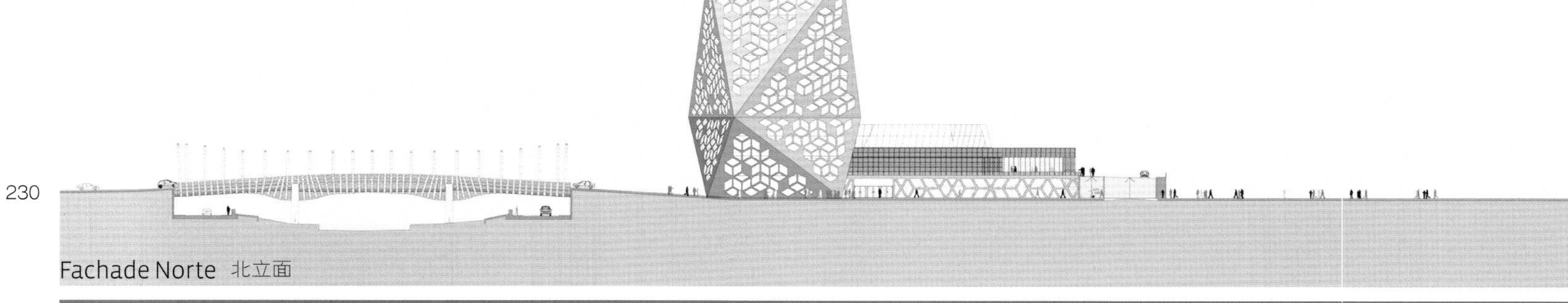

Fachade Norte 北立面

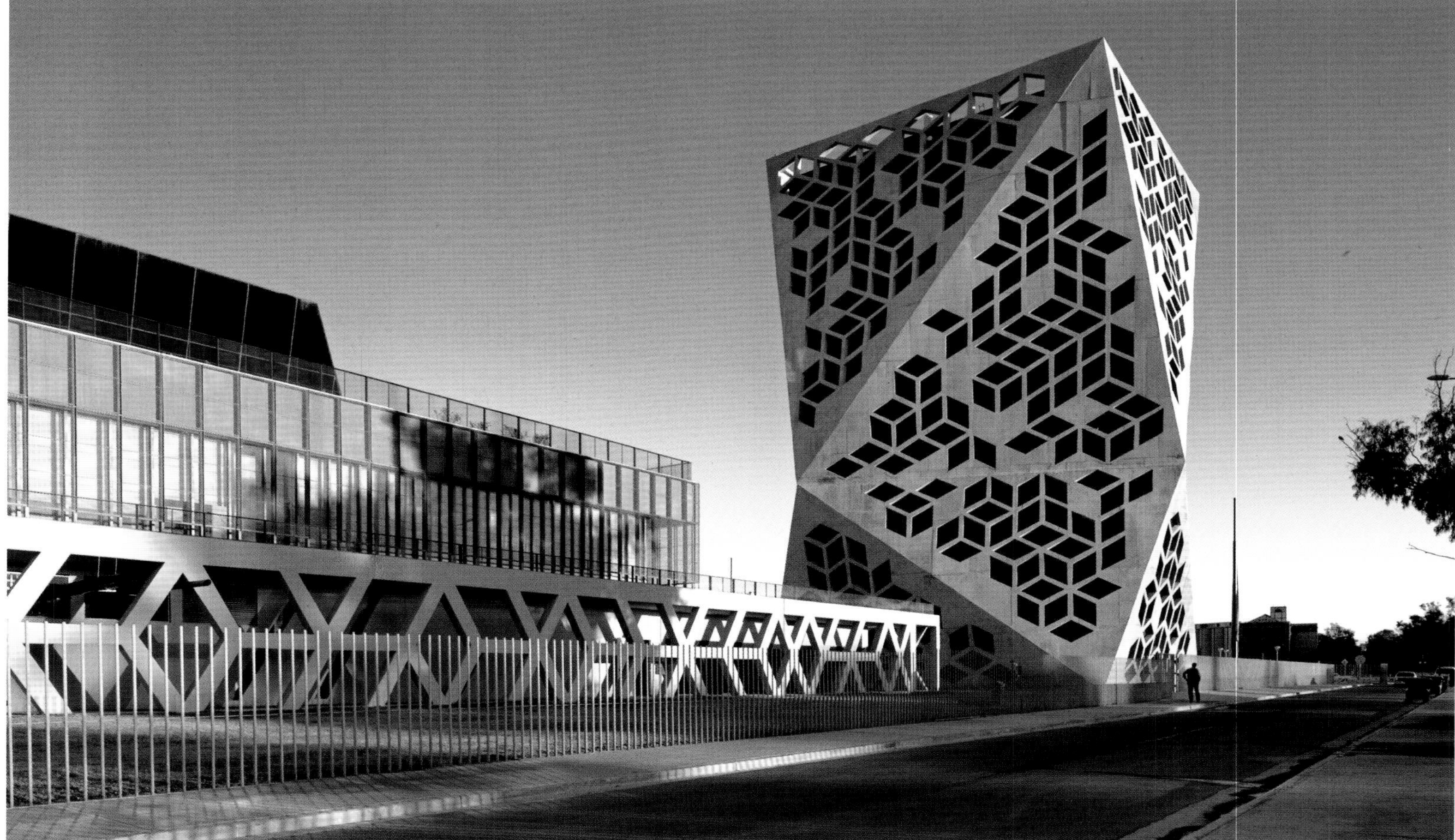

The regular rhombus framework is orderly arranged, making the architecture outline novel, striking and powerful. Specifically, the square glass component makes the whole building seem to be in high spirits and meanwhile gives visionary yet realistic experience.

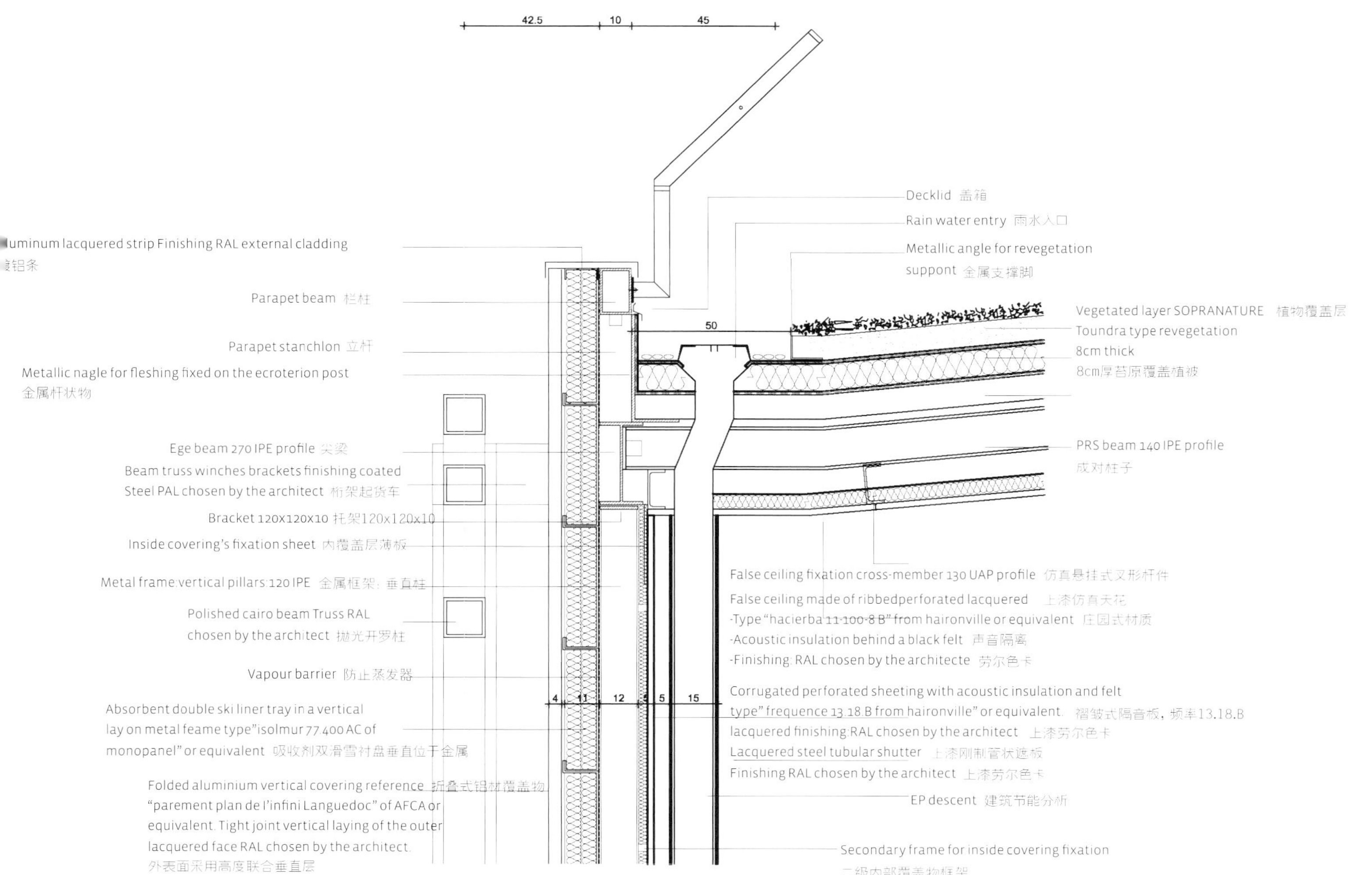

Detail facade pignon 立面细部

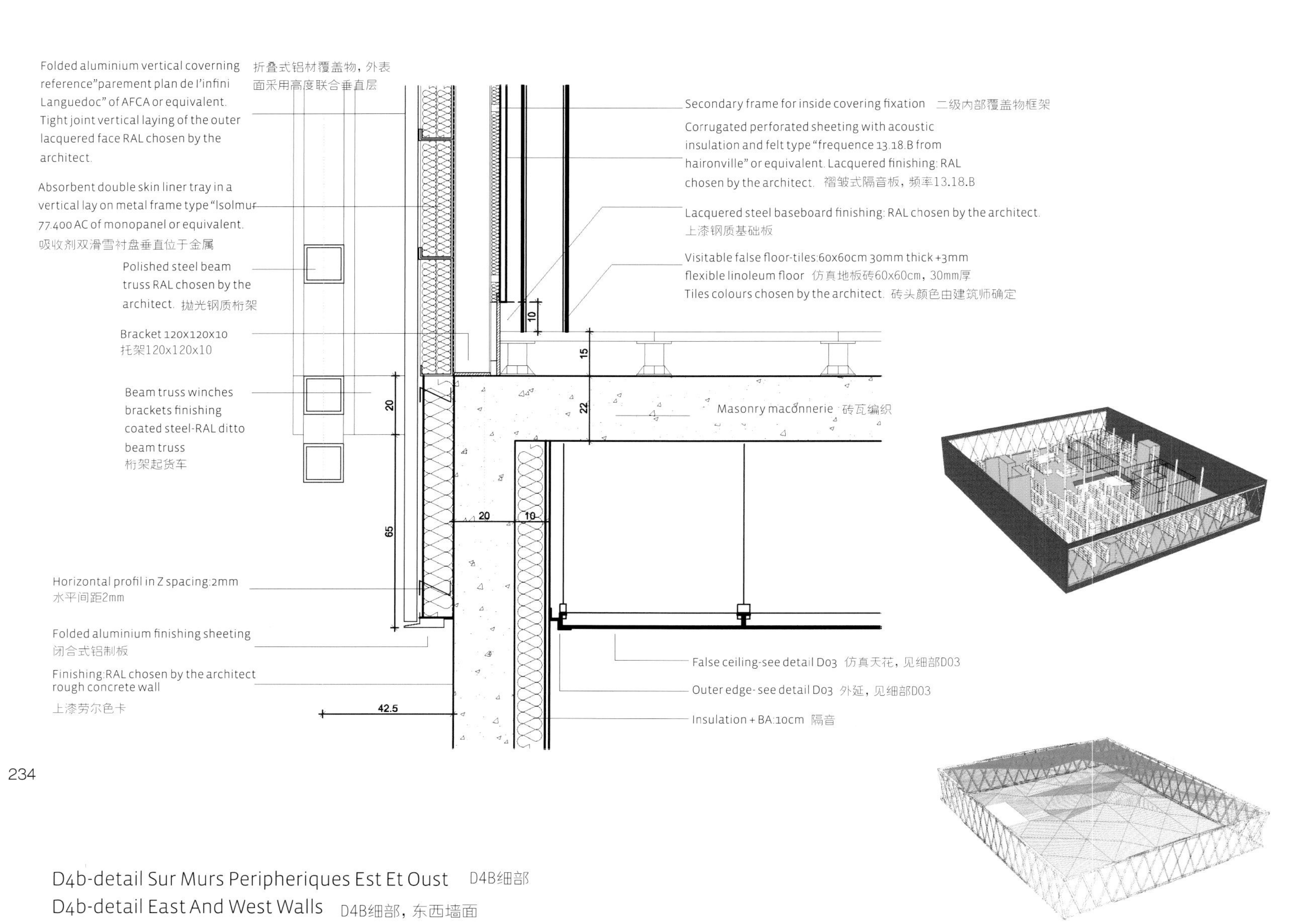

D4b-detail Sur Murs Peripheriques Est Et Oust D4B细部
D4b-detail East And West Walls D4B细部，东西墙面

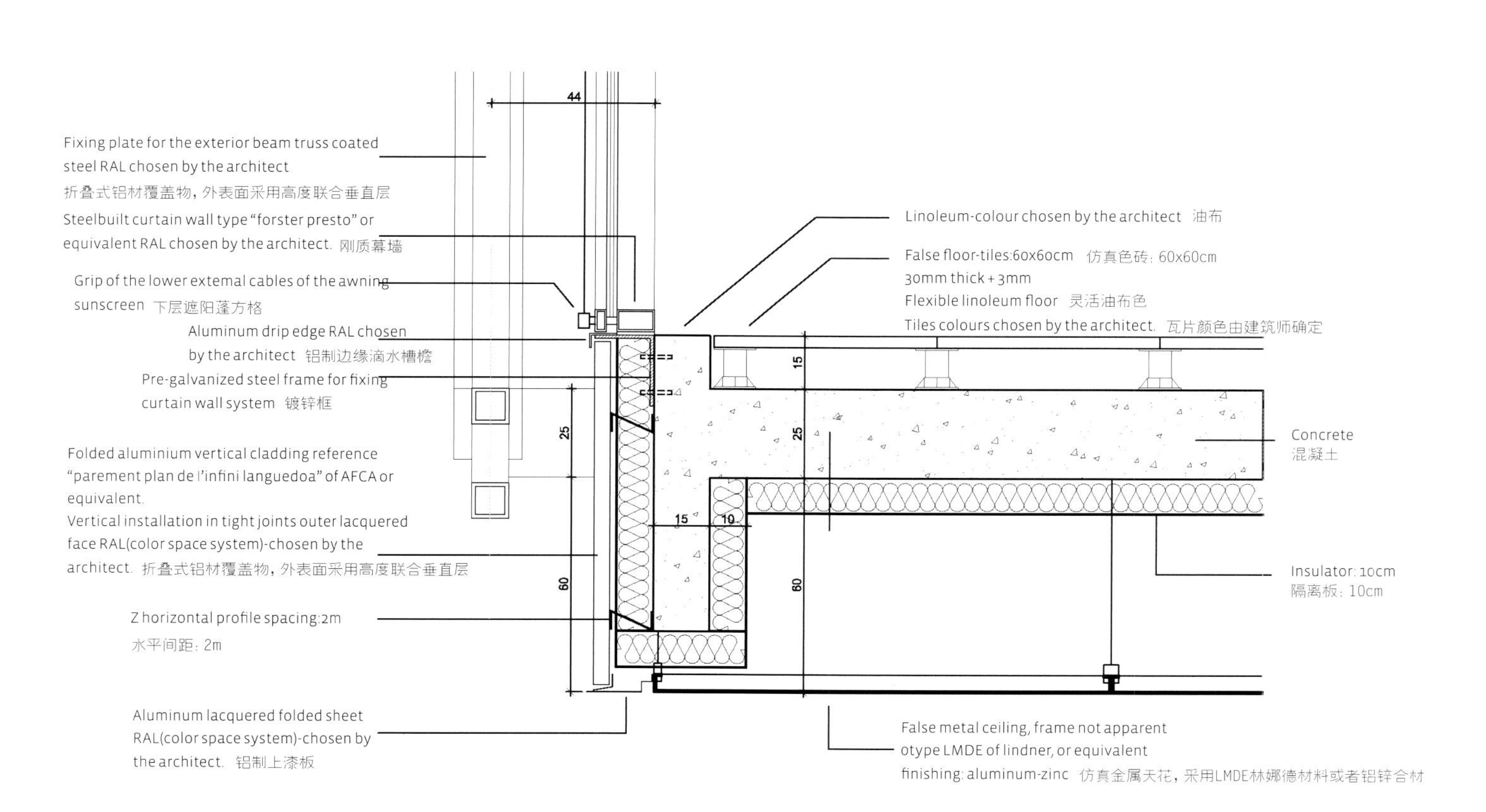

Detail façade vitree 立面细部
Detail glazed façade 玻璃立面细部

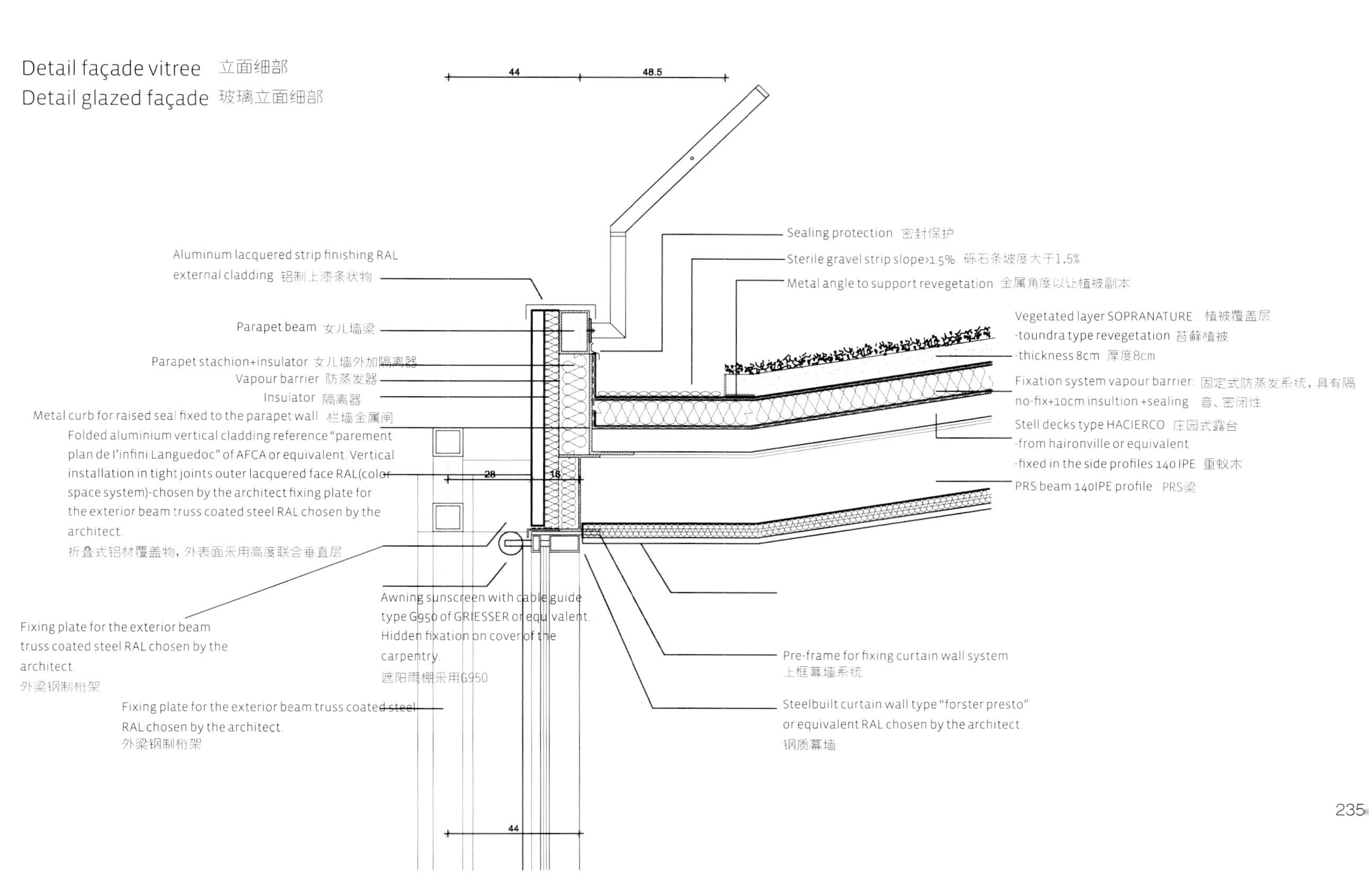

规则的菱形外框架依次有节奏地排列，使建筑外形新颖别致，极为醒目而有力度感。作为方形玻璃体的外部构图，它使整个建筑物都显得格外精神，沉浸在几何形态所带来的韵律美，给人虚幻而又实在的视觉体验。

Distortion is the twisted mutation of geometry based upon certain regular patterns and yields novel plane and space relations, giving a sculpture-like façade image. The inspiration of twisted shape comes from "fence" which is made of several rhombuses, delivering a new concept of integrating space excitement and abstract, creating an irregular dynamic. Meanwhile, its unique yet natural geometry connects with nature, like a inhabitable sculpture, showing special peace and strong 3-D effect.

扭曲是将几何平面按某种规律进行不同角度的扭曲变异，在力的作用下整体变化而产生出奇妙的新型平面空间关系，它赋予建筑雕塑般的立面形象。这种卷曲的造型灵感来自现实中"篱笆"，它由多个菱形构成，传达了一个新型的概念，即将先进世界的空间刺激和抽象空间的图式清晰地合二为一，旨在营造一个不规则的运动的感觉。同时，它又以其独特而又自然的几何形态取得了与自然的联系，仿佛是"可以居住的雕塑"一样，流露出特有的静谧表情和强烈的立体形态。

Scetch 01: ...at the fence
草图1：栅栏处

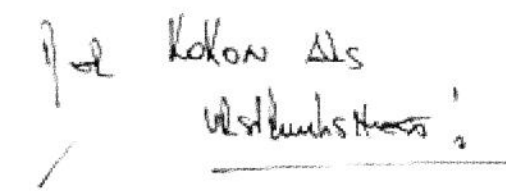

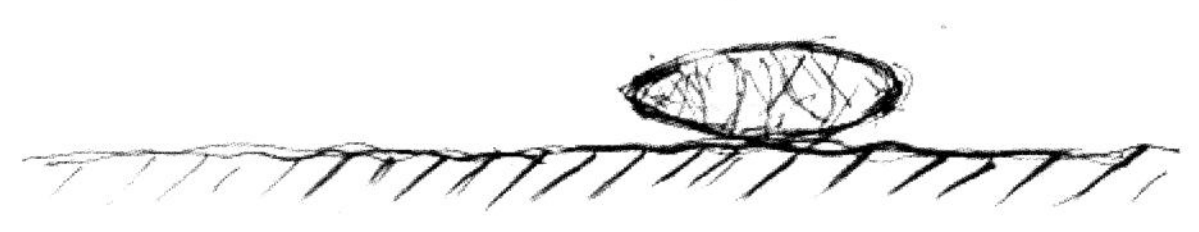

Scetch 02: the cocoon...closed...opening
草图2：塑料披盖

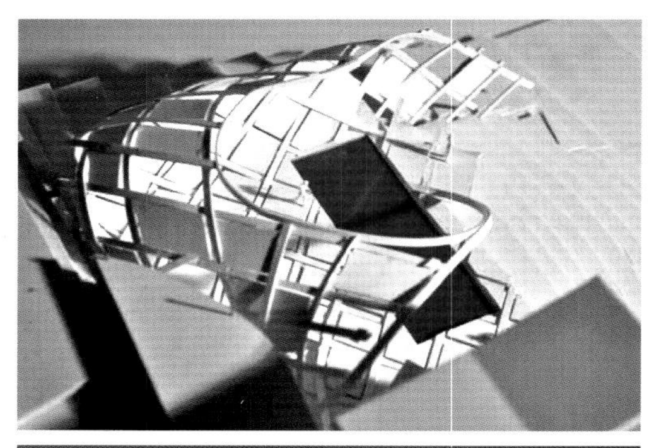

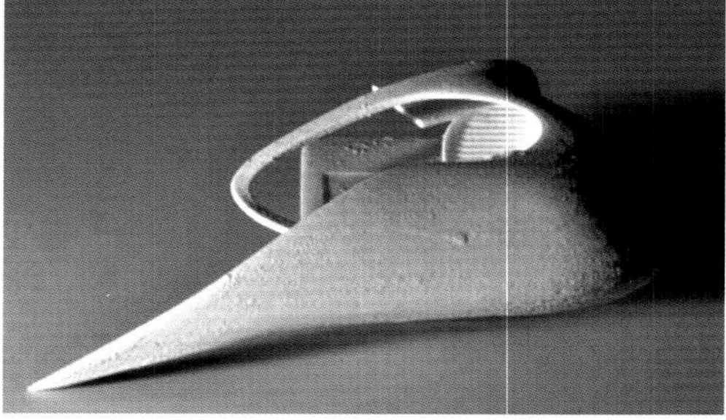

Model 01: studies of form 模型1：形式研究

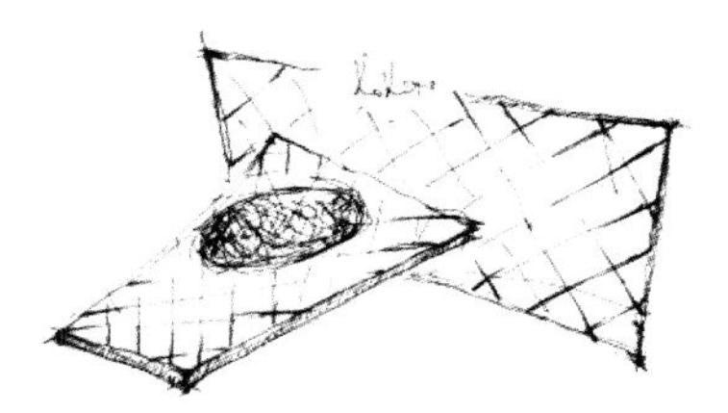

Scetch 03: we start with an orthogonal fence
草图3：采用垂直栅栏

Concept: the fence
概念：栅栏

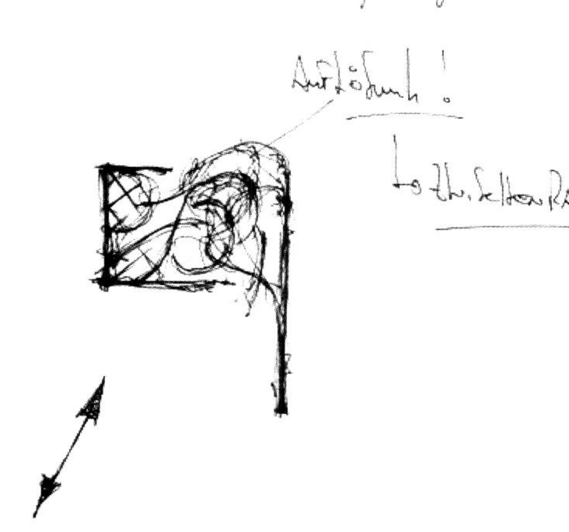

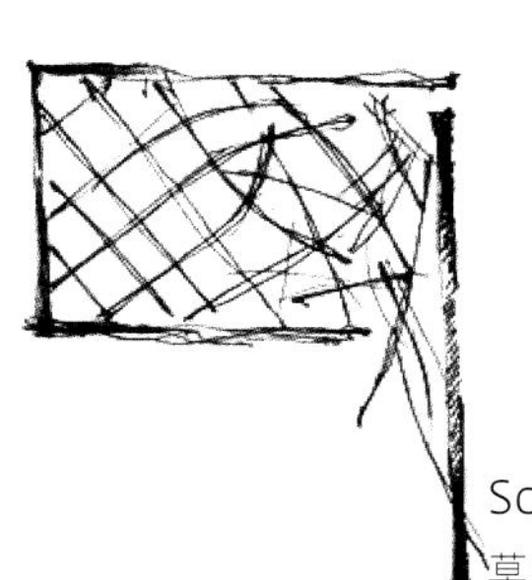

Scetch 04:der the fence describes the spac
草图4：栅栏描述空间

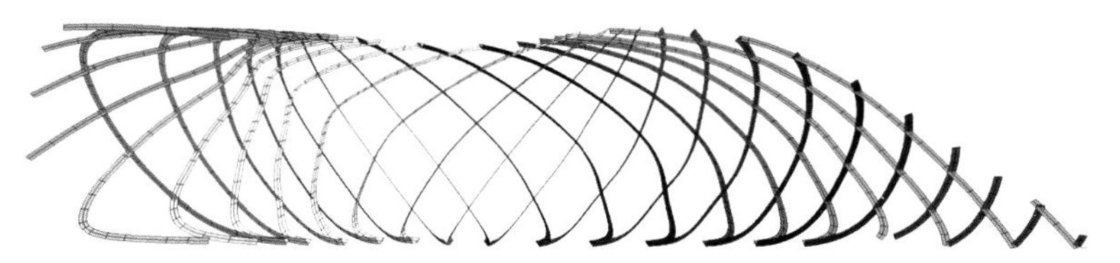

Concept of the construction 建设理念

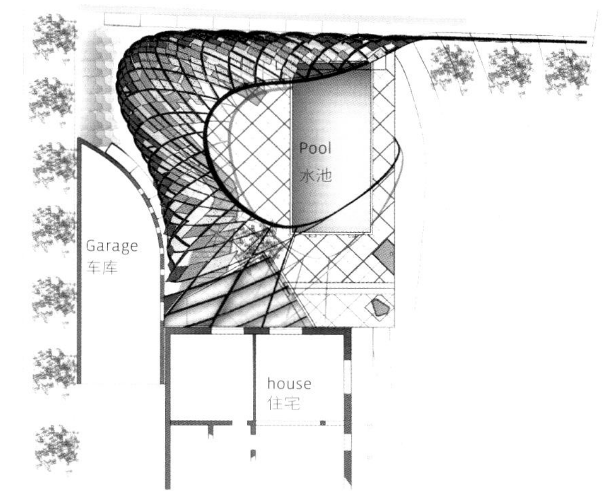

Ground Floor 首层

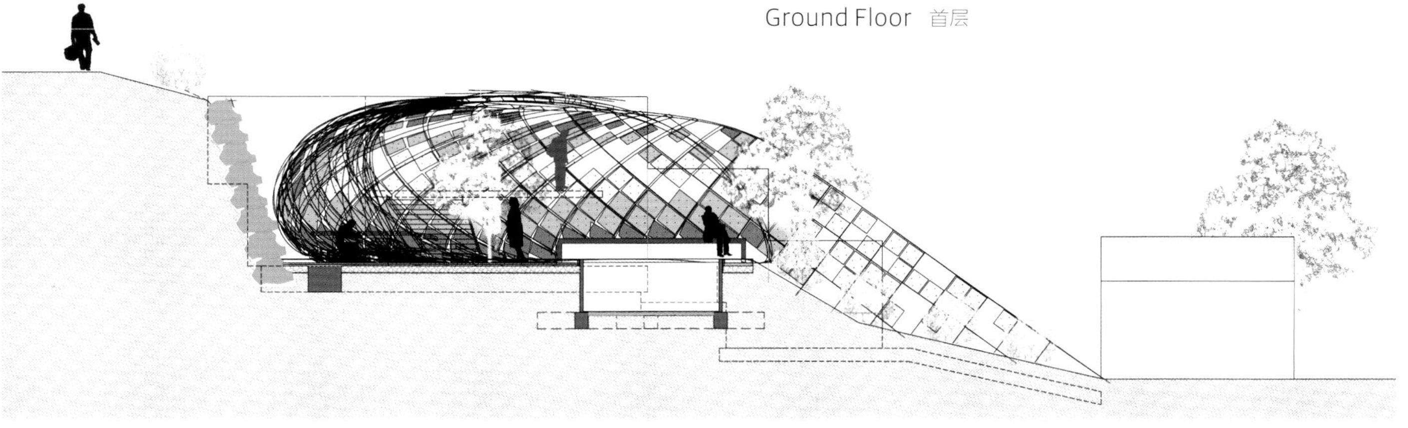

Section 剖面图

Model 02: concept
单体分解

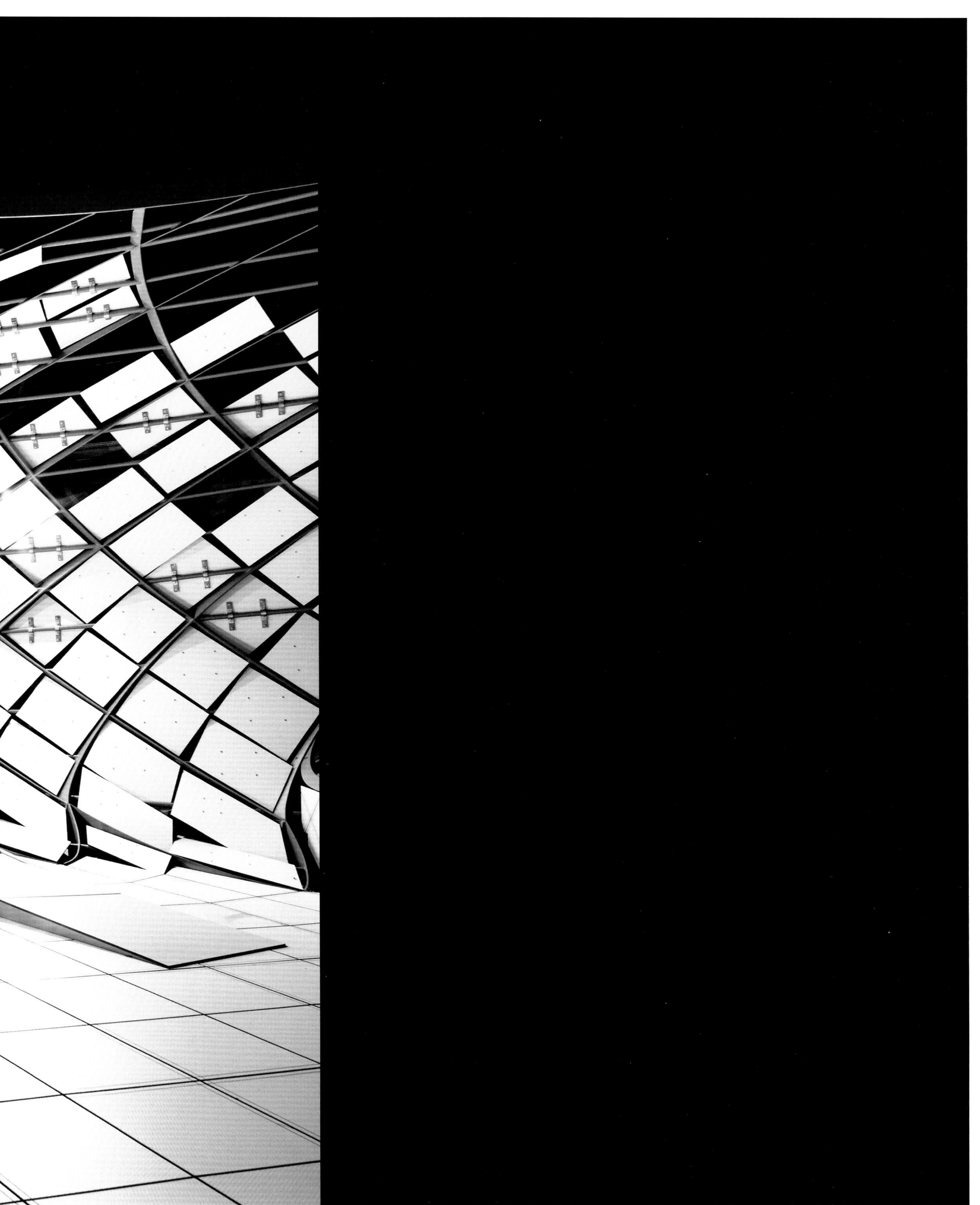

INTEGRATED, HARMONIOUS, REGULAR, ETERNAL

圆融和谐，生命的轮回，规则与永恒的象征

A circle
is the shape of the universe and it gives
people stable mind and helps them recall the life circle.
Everything was born in the universe, then grows up, vanishes
and then grows up and perishes, and on. A circle without beginning or end
represents the pure, flowing and dense geometry form. A circle is the symbol of unity
and its flexibility allows itself to be scaled to any diameter, creating the regular and eternal
beauty.

For buildings with the round façade, they withstand little wind force due to the circle structure. Round buildings are both inward and outward, with the circulating beauty and are more stable. Thus round buildings are better at anti-seismic abilities.

Due to the rotating symmetry, a circle is considered the most beautiful figure in the world and it displays a complete symmetry and harmony, coordinating the whole and the parts. The symmetry gives circles an unchanged curvature. Obviously, the shape makes people feel perfect and steady.

In addition, a circle is more self-satisfied. It is a thing owned solely by Gods. It is the symbol of the world system. Its circumference rotates around the axis, singing unuttered songs and this is its god feature. Deep down into a circle, it bursts out unparalleled divinity glory. The round shape is associated with the sun, the moon, the sky, the universe, the female sex organ, etc.

圆形是圆满的形状，是宇宙的形态。它能给人衡定的心态，使人们去想象生命的轮回。一切的生命都孕育在宇宙中，出生、成长、毁灭、再重生，再毁灭，轮回不断，生生不息。圆形，没有开始，没有结束，代表着至纯、流动和紧凑的几何形态。圆也是团结的象征，它的灵活多变可以允许被任意挤压到精确比例和直径，创造出规则和永恒的美感。

对于圆形外立面的建筑，由于其圆弧形的外部构造，受力面很小，风力很易分散，承受的风力就会变小。圆形建筑既具有极强的内向性，也具有极佳的外向性，向内聚于中心点，向外可以均匀发散，不仅充满循环运动之美，而且其房屋结构构件的整体链接会更加地牢固、地基会更加稳定。因此，在抗震性方面，圆形建筑比方形建筑会更好一些。

圆形之所以被人们认为是平面图形中最美的图形，那是因为圆具有完全转动的对称性，显示出一种绝对的对称与和谐，使整体与部分变得十分协调，因而完美而神圣。这种过强的对称性使圆形具有一种不变的曲率，它由圆形仅有的一种结构条件所决定，即“圆形轨道上的点离中心点的距离相等”，显然，这种图形使人们感到完美无缺、稳定凝重。

此外，圆更是最高的自足和圆满。它是被神所垄断的事物。圆是世界体系的象征。圆拥有隐秘的轴心，
圆周环绕它运转，唱出无言的礼赞，而这就是圆的一个神教本性。从圆的形体深处，迸发出
了无与伦比的神学光芒。 圆形造型支持了对太阳、月亮、天空和宇宙、母性生
殖器等神性事物的联想，并成为它们的天然能指。

Unique Oval Crossing Shape – The unique structure ensures that all balconies with a two-story height introduce natural light as much as possible. Apart from the benched structure, the whole building is orderly distributed in a right angle. Its design echoes with coast wetland and it is the specific feature of Australian landscape.

独特的蛋形交叉结构——这个与众不同的结构保证了所有的阳台都有双层高度，尽可能地引入自然光照。除了弯曲的部分，整座建筑的布局相当规整，呈直角形。设计的几何形状与湾口湿地的弯曲形状相对应，这是海岸潮汐冲刷的结果，也是澳大利亚景观的一个独特之处。

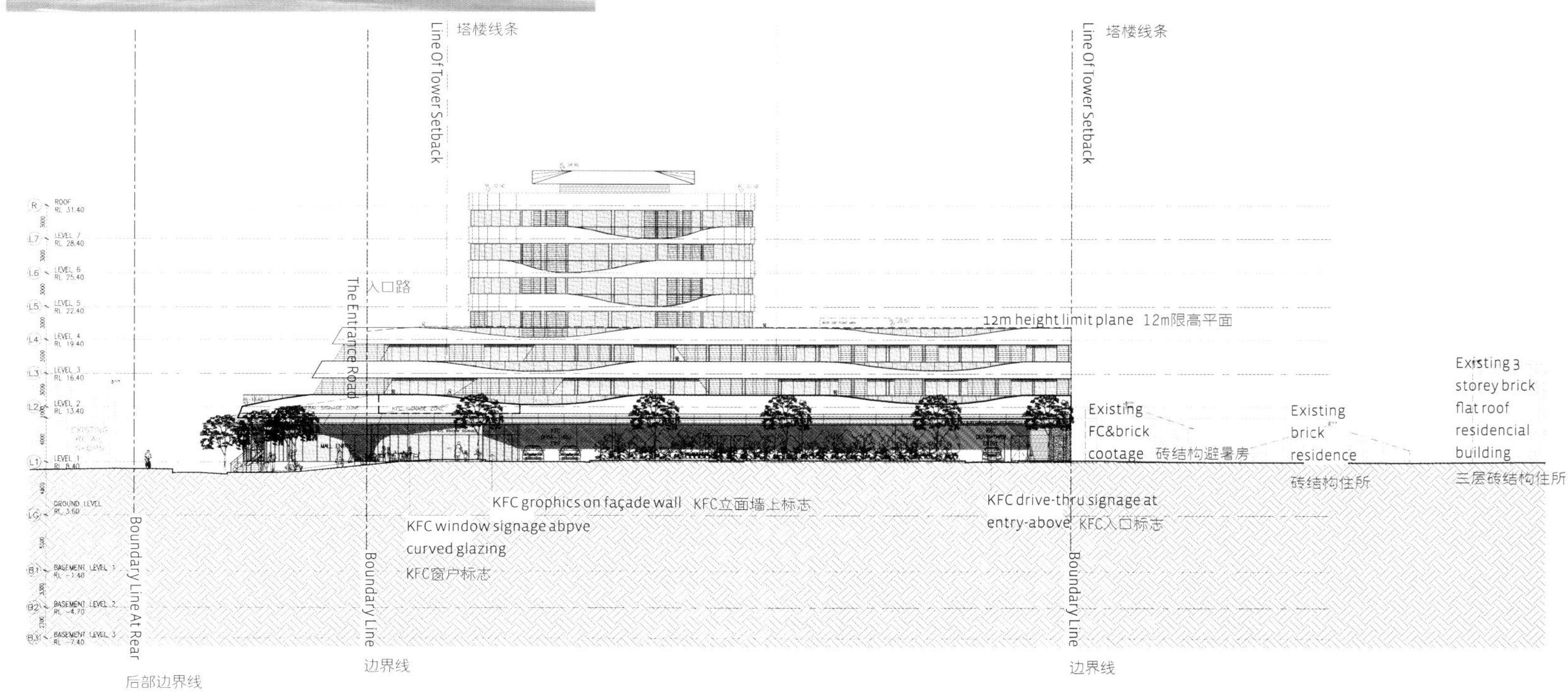

The inspiration of the project comes from Beijing and its five flowing round shapes join at the bridge and harmonize with each other, forming a fluid combination where fluid curves makes the building gentle and the whole building exudes a very striking and powerful breath.

该项目设计灵感来自规模宏大的北京，5个连续流动的圆形的建筑形体通过桥梁连接在一起，彼此协调，成为一个无死角的流动性组合。在这里，流畅的圆弧形线条使建筑不再是刚性的，而是柔性的、流动的，整体建筑拥有鲜明而强烈的气场。

DIVERSE, VIVID FACADE

形态多样，灵动自由的界面形象

The geometry in architecture is abstract, practical, one-sided and dull. Although the single geometry form is easy to take on, it is often too simple and mechanical and can't reflect the complex texture of architecture. For most buildings, the building with the single form is rarely seen. Many buildings are in mixed forms, like square with trapezoid, line with square, circle with irregular shape, etc. Different geometry combinations provide buildings with diverse façade images.

Mixed form tends to reflect the diverse, complete and abstract image of architecture. Buildings of mix form are mostly composed of square, round, polygon, trapezoid, etc. Due to diversity and irregularity of form, the building takes on different images which strengthen the relations with the surroundings, reaching a higher level of sympathy.

The mix of different planes makes the façade more diverse and brings a powerful space sense. Set off by the mix planes and light, the building forms a strong contrast between "bright" and "dim", taking on an extensive volume and grand soul, with much vigor. It breaks only the dullness of boxes and employs plane variation to integrate with functions and environmental elements. Architecture is not just the wall of closure and it has been redefined through the mixed geometry to be an architectural element with a distance from function.

作为建筑的高度抽象，单一的几何形有其实用的一面，也有某种片面性和单调性，单一的几何形构成的结构合乎理性且易于设计和建造，但所产生的形式往往有些过于简洁和机械，难以反映具有复杂组织肌理的建筑形态。在大多数建筑中，纯粹单一的几何形态的建筑是很少见的，大量的都是它们的组合体，如方形与梯形、线形与方形、圆形与不规划形，等等。不同种类的几何形的平面组合，能为建筑提供丰富多样的立面形象。

混合形更易于反映出建筑多变、完整、抽象的形象。混合形多由基本形中的方形、圆形、多边形、梯形等构成，由于形态的多样化和不规整，建筑形体也表现出多变、灵活的构图形式。这种灵活的平面构成形式，加强了建筑与环境的沟通，使构筑的建筑形象能与自然相互融合，让建筑与自然达成一种深层次的共鸣。

图形与图形的混合，除了使立面形象更趋于丰富和灵活外，更带给人一种强烈的空间意向，建筑在混合形平面的衬托与光的渲染下，形成明与暗，虚与实的强烈对比，呈现出极富扩张力的体积感和凝重而博大的气魄，由内而外散发出一股源源不息的生命力。它打破了只是方盒子的单调性，造型上采用不同几何形之间的穿插、叠合、重复、凹凸形成多样的平面变化，再依据功能或环境因素进行组合。建筑已不再是简单的墙体围合，而通过混合的几何形被重新定义，界面成为了一个与功能保持一定距离的具有相当自由度的建筑语素。

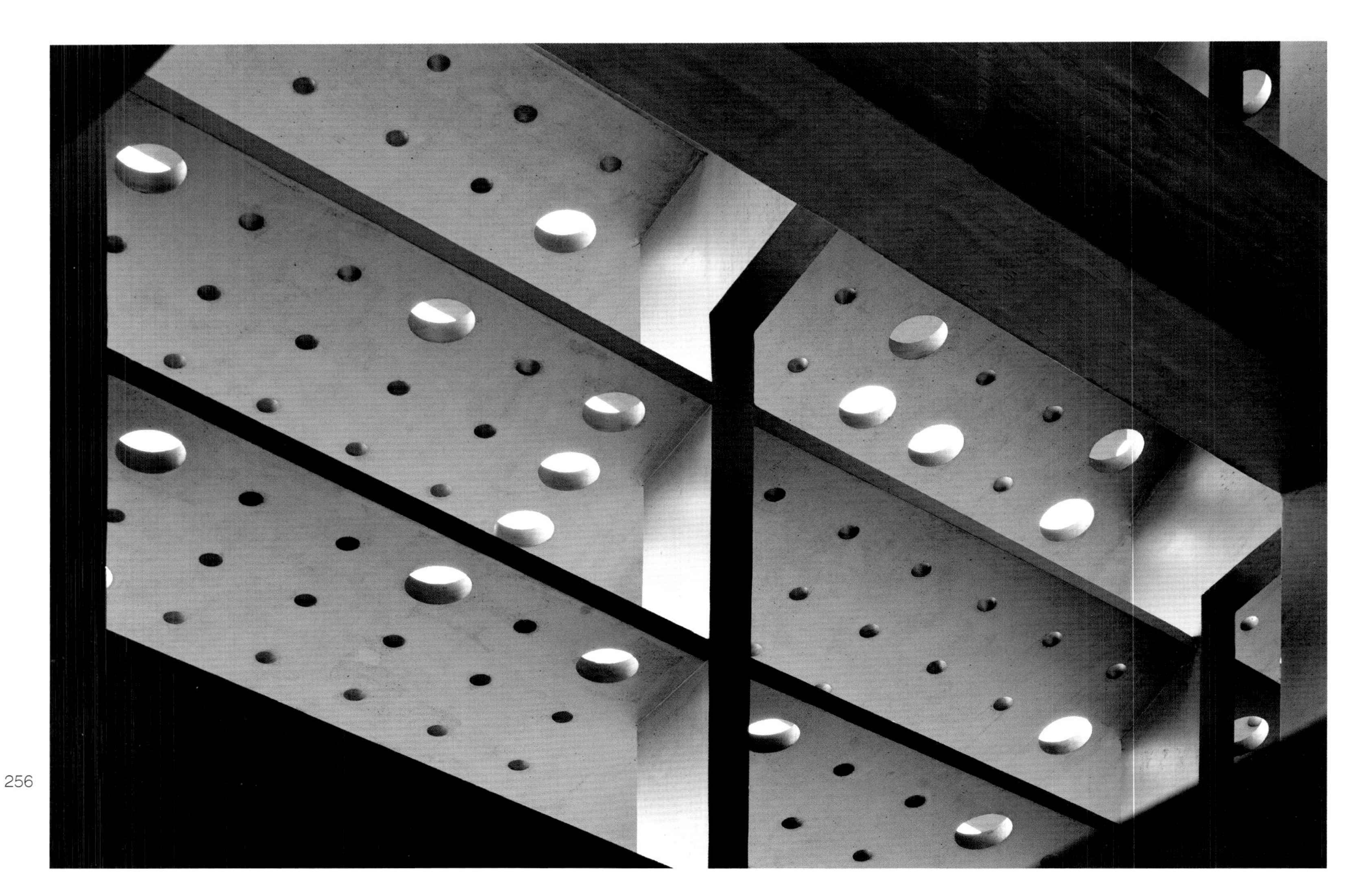

Chimney effect sucks hot and humid air out of the building
烟囱将热且潮空气鼓出建筑

Roof 屋顶 +4768
Roof garden 屋顶花园 +4488
Residence / gym 住所/体育馆 +4125
Offices / design studios 办公室/设计工作室 +3762
+3399
+3036
+2673
+2310
+1947
+1584
+1221
Level 3-10 factory floors 工厂3~10层 +858
Cafeteria 咖啡屋 +495
Showroom / storage 展览室/贮藏室 +132
Street level 街边 +0
Parking 停车 -346.5

Clear floor space 楼层空间
Service core 服务核心
Service / fire lane 服务/消防道
Vehicle / bike lane 机动车辆/自行车道

Cross-section
横向剖面图

The design is to building an open, continual, luminous and ventilated work space. The façade of the 11-story building adopts the sunshade to reflect the aesthetic beauty and more importantly to bring about the functions of lighting, sunshade and ventilation.

设计的目标在于建立一个开放的、连续的、光线充足、通风良好的工作空间。这个11层的建筑外墙采用的遮阳鳍片，除了体现建筑的几何美感外，更重要的，线形与圆形相交织的外表皮，实现了采光、遮阳与通风的所有功能。

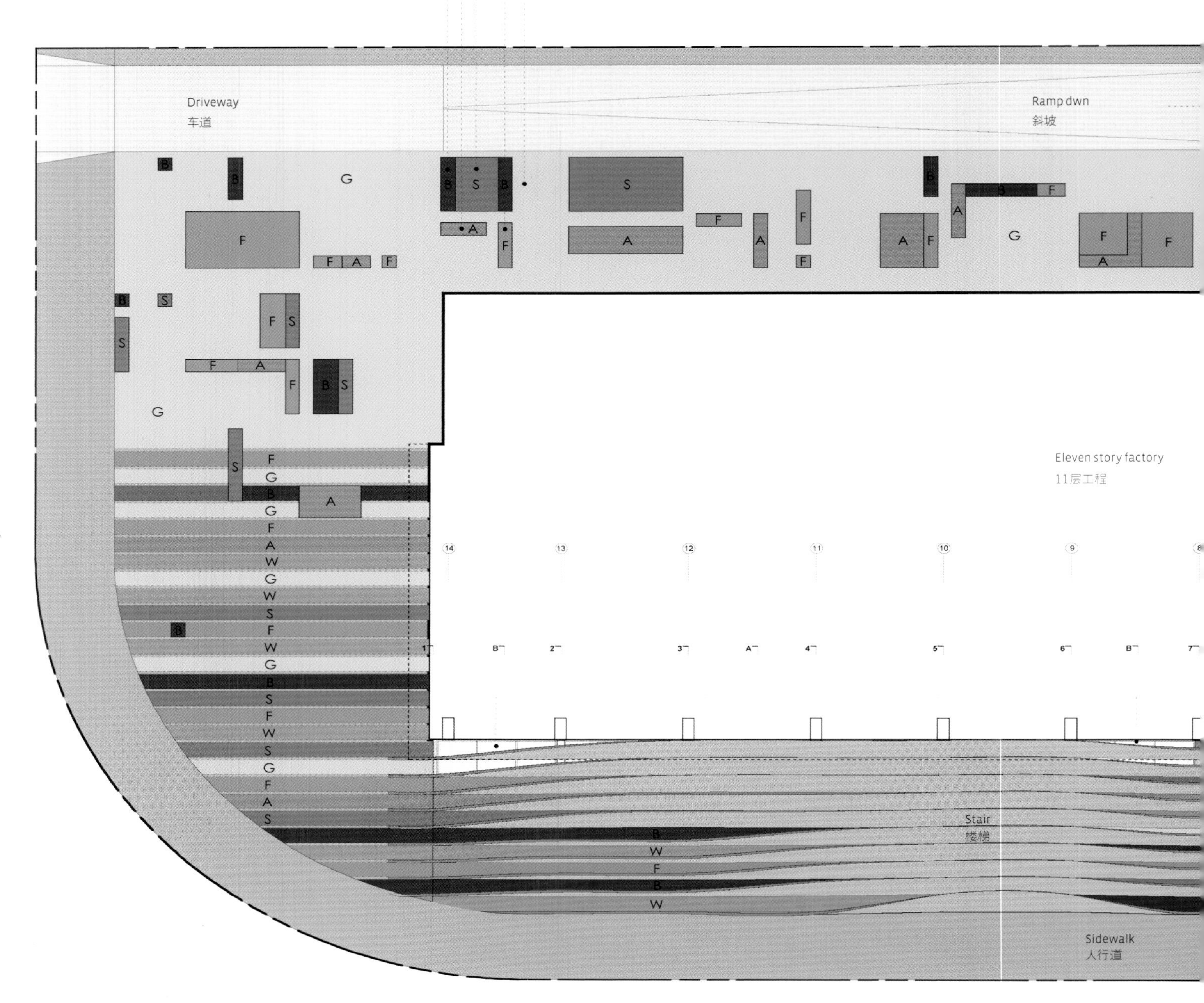

Site plan-landscape proposal
景观总平面图方案

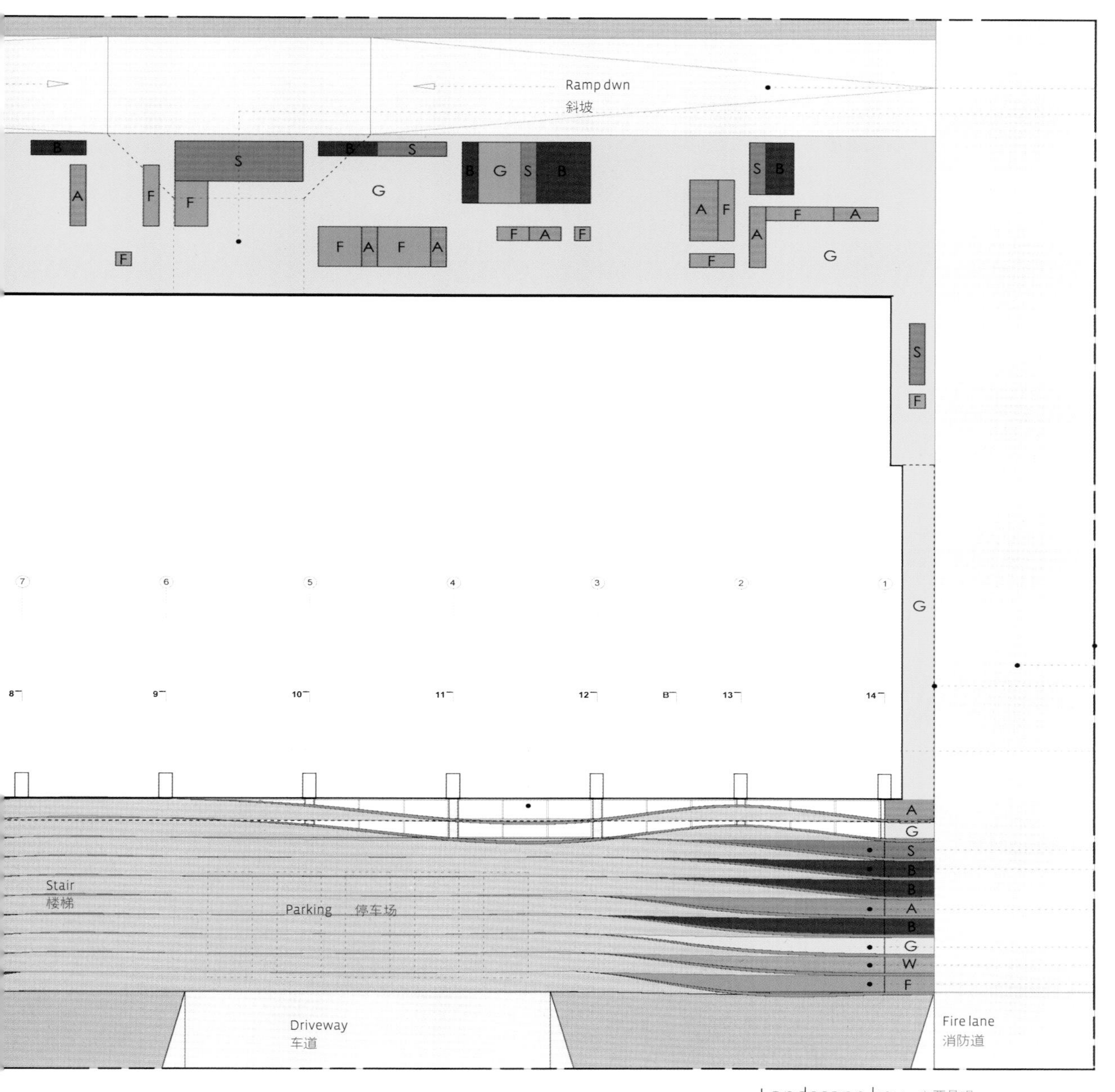

Car / bicycle exit
车出口

Basement parking entry and exit below
地下停车场出入口

No bamboo or shrubs located in this area to maintain fire engine lane. Only gravel and flowers to occupty this area.
消防道无竹子或灌木，只有砂砾和花

Lot line 基地线
Fire engine lane 消防道
Edge of fins above
散热片

Skylights below
天窗

Shrubs 灌木
Bamboo
竹子
Agate stone 玛瑙石

Grass 草
Water 水
Flower 花

Landscape key 主要景观

Height 高度	Color 颜色	Title 标题	ABBR. 缩写
300 CM		Bamboo 竹子	(B)
90 CM		Shrubs 灌木	(S)
45 CM		Flowers 花	(F)
10 CM		Grass 草	(G)
5 CM		Agate stone 玛瑙石	(A)
0 CM		Water 水	(W)

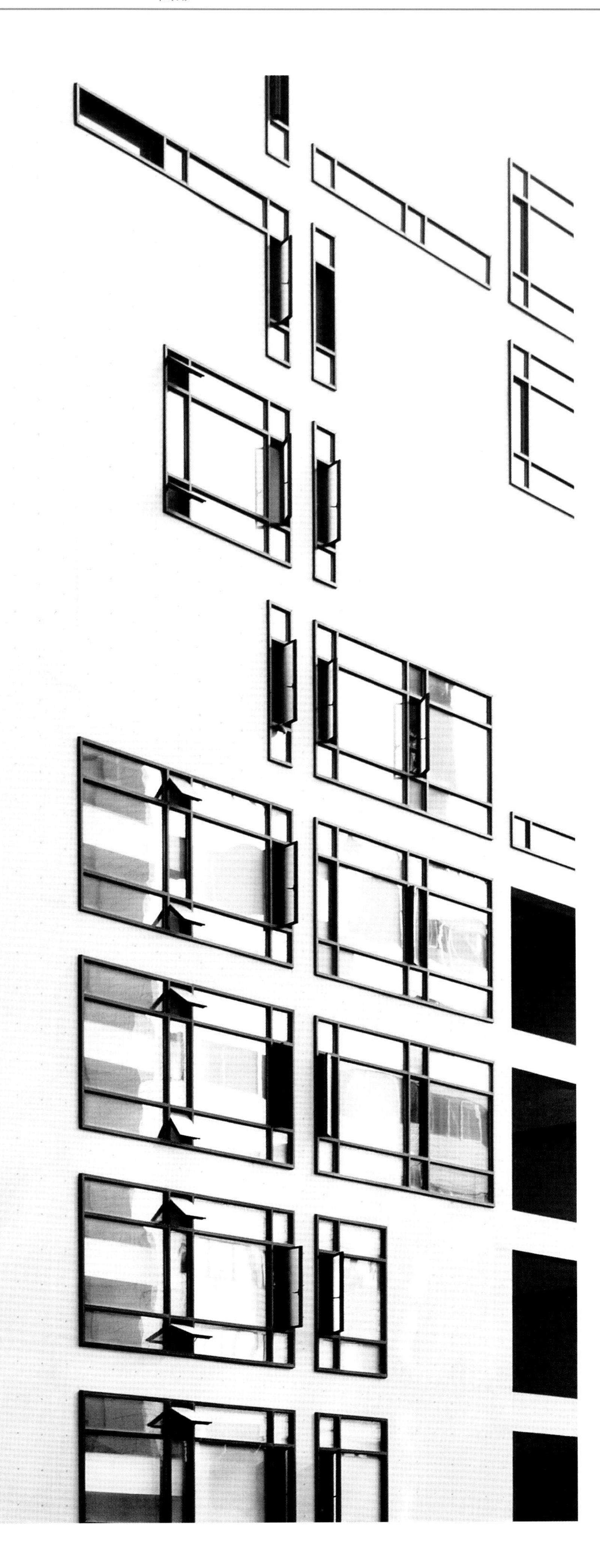

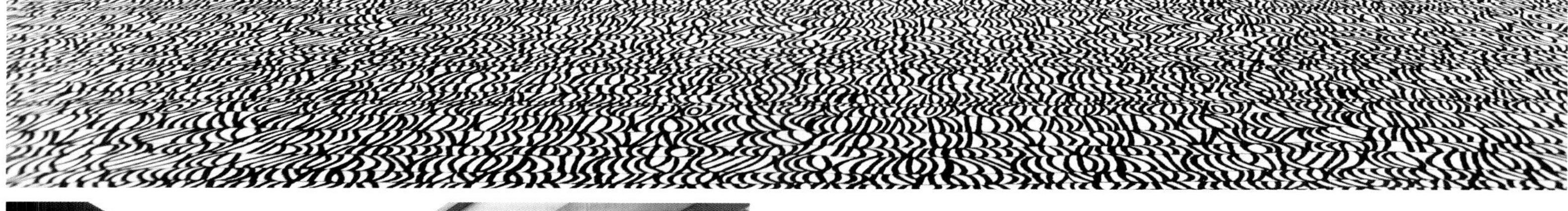

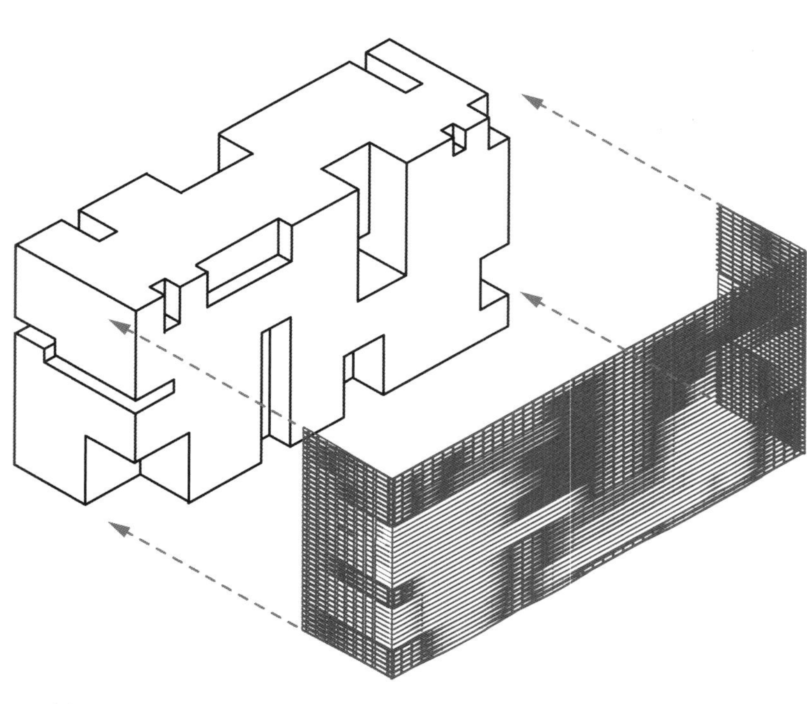

Chimney Diagram

烟囱图

ICBC
中国工商银行

中国工商银行

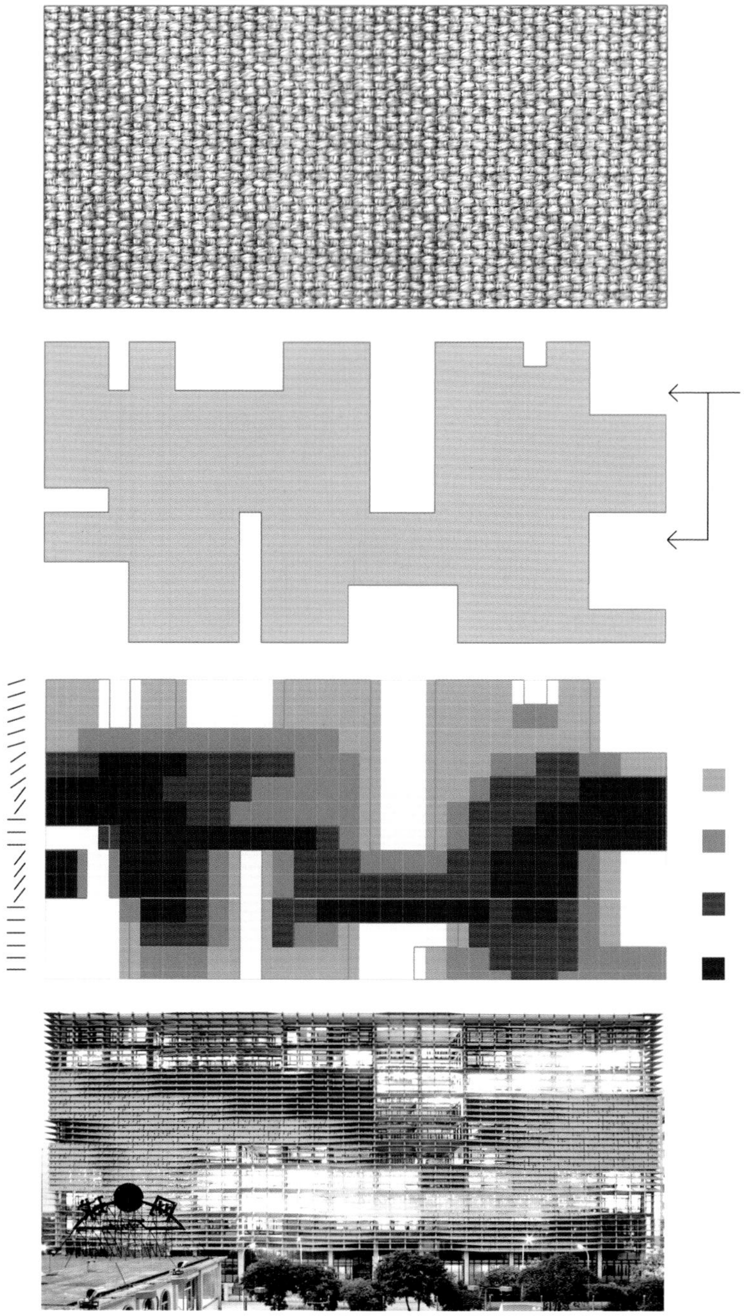

South Facade 南立面

1

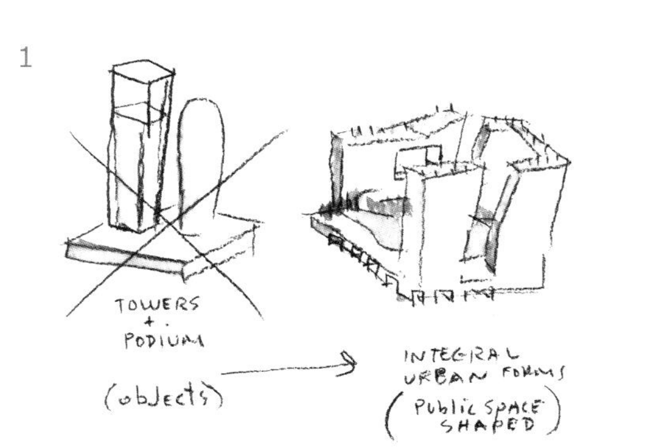

2

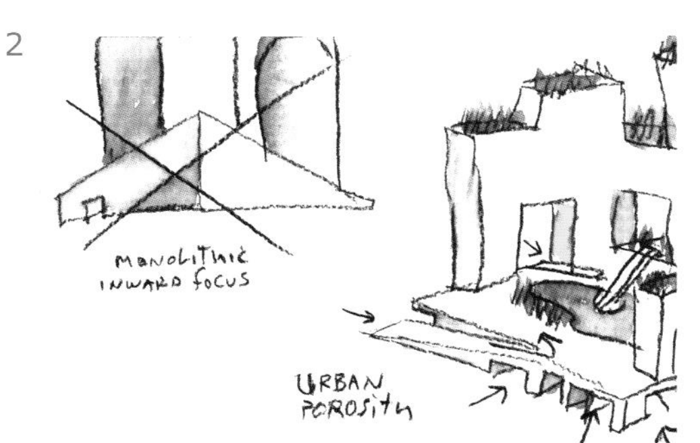

1. Integral Urban Functions Shape Public Space 现代一体化公共区域
2. Porosity 多孔
3. Microurbanism 微现代主义
4. Super-green Architecture 超级绿色建筑
5. "Three Valleys" inner Gardens "三峡谷"内花园
6. Spatial Geometry Lit Via Pond Skylights 池塘天窗照亮空间几何形

3

4

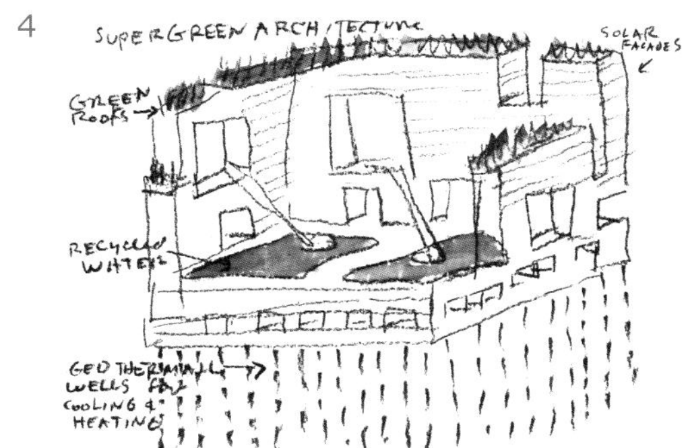

5

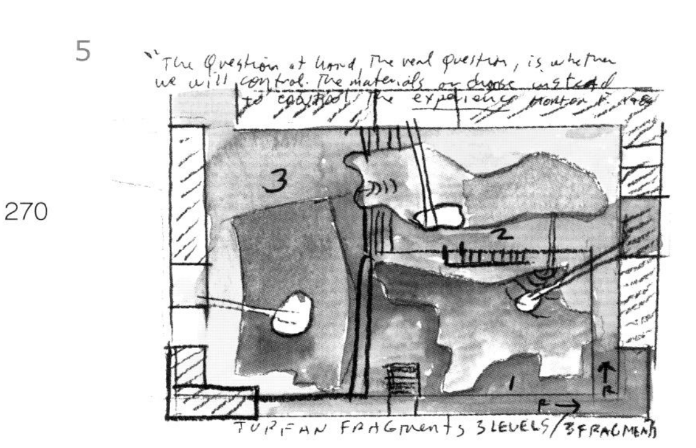

6

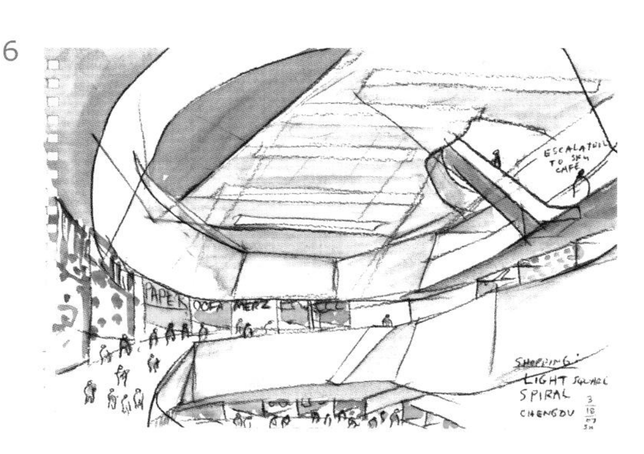

"Cut bubble building" – The whole building is in a hung, piercing and irregular shape. Its façade displays the seemingly bulky yet slim beauty and realizes the aesthetic level of bringing excitement to plainness. The whole building is like a bubble cut into five tower buildings and form the unique façade structure. To ensure every building is shined in the sun for two hours, it is very slim in design and the building exterior line conforms to the sunshine position requirement. Thus the building is "cut". The exceptional from head to toe expresses the art power.

"切开的泡沫块"——整个建筑呈大悬挑、大孔洞和不规则倾斜状。外立面体现了"大巧若拙"，实现了把"绚烂之极归于平淡"的超高审美境界。整体结构宛如泡沫块一般，切割出5栋塔楼，形成独特的立面结构。为了保证每栋建筑物每天至少能有2小时的日照，每栋楼都设计得很"薄"，并且建筑物外部线条完全契合这一地点的日照角度，所以说它是"切开的"。这座特立独行的建筑从外观到内涵，都表现出艺术的生命力。

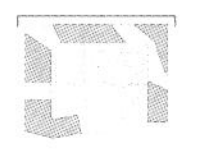

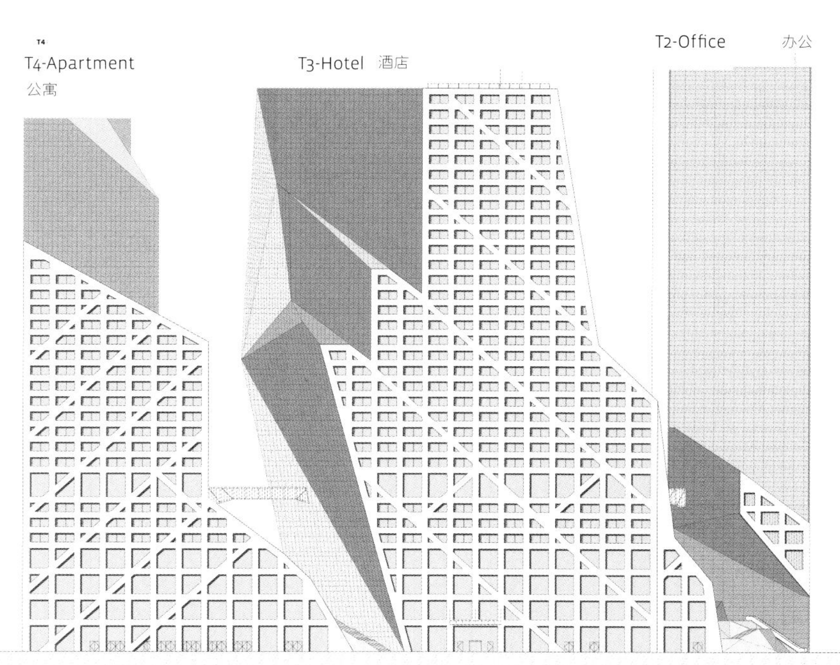

East Elevation 东立面

成都来福士广场封顶仪式
Raffles City Chengdu Topping Out Ceremony
CapitaLand
Raffles City

CapitaLand
AMD

新世纪电脑商城
信德实业

The irregular building above water adopts a façade with curved square blocks to form a unique plane form, with a slanting volume, making the whole building flexible and interesting.

这个被置于水上的不规则建筑体，外立面利用被串起的有弧度的方形块构成了独特的平面形态，倾斜的量体，使整个建筑显得灵活而有趣。

The small building covered in bronze façade is located near Sydney and it looks like an open roof. This concept comes from the regulation that new buildings must adopt the traditional sloping roof.

The rotten bronze materials form the exterior wall and roof and blend with the building. The undulating roof meets the requirement. Its form is inspired by the triangular four-slope roof. Its southern base is adjacent to the surrounding building to ensure privacy. The windows on the northern façade can enjoy the plants in the garden.

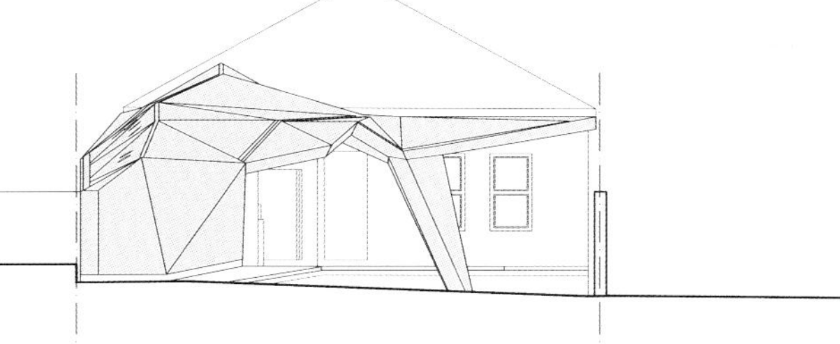

South East Elevation 西南立面

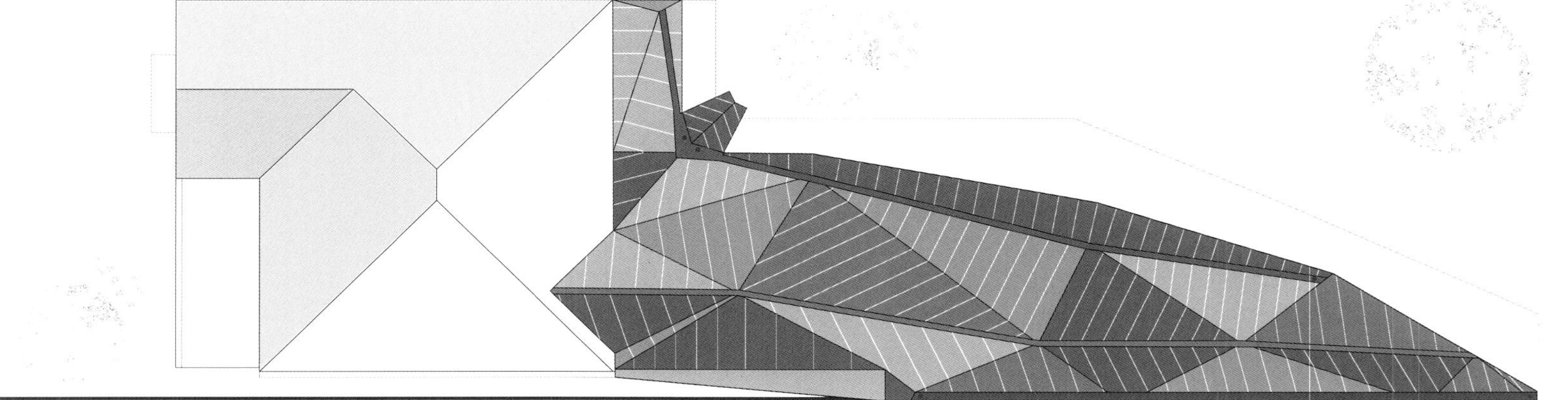

Roof Plan 屋顶平面图

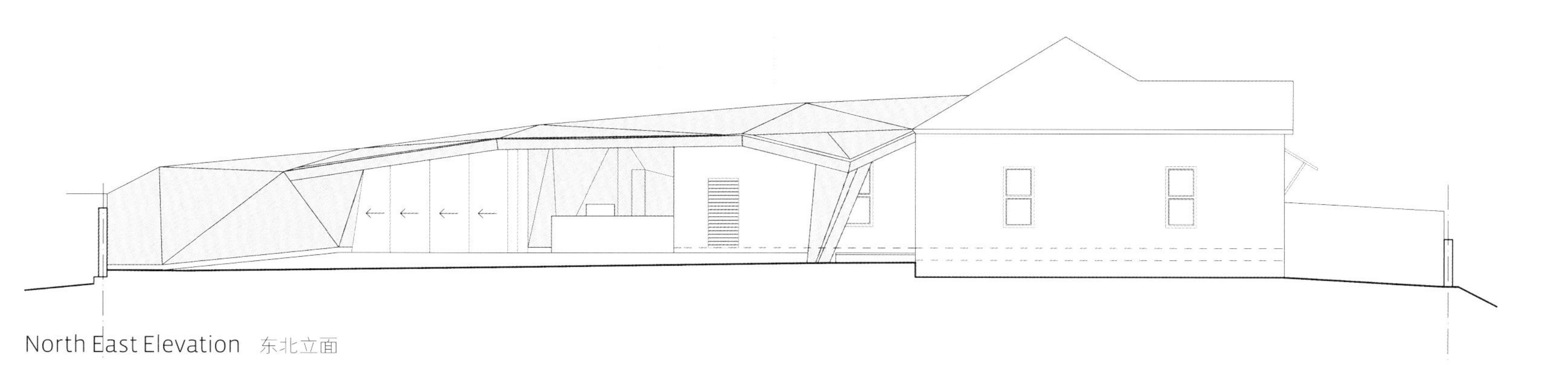

North East Elevation 东北立面

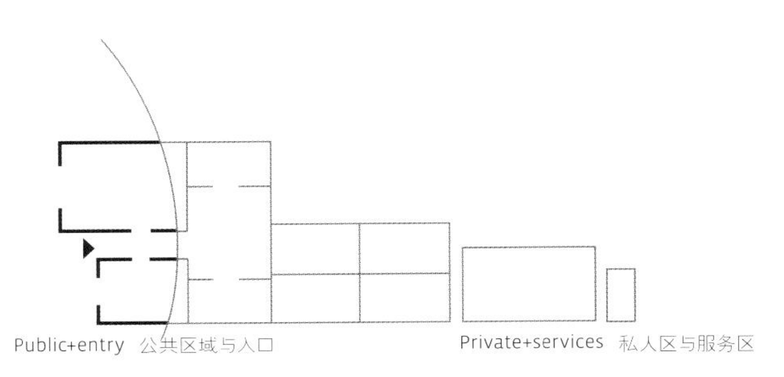

Historical usage

历史用法

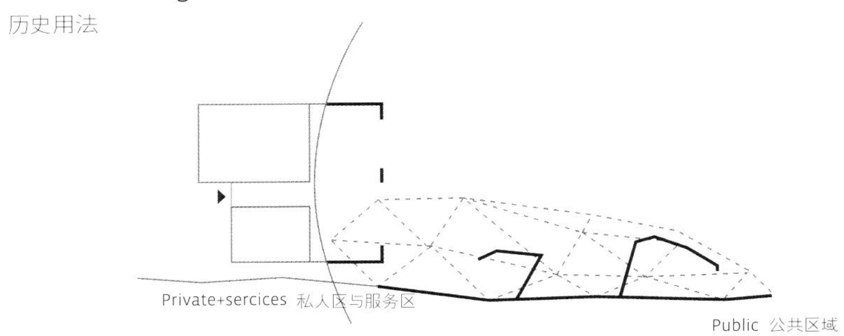

Contemporary usage

现代用法

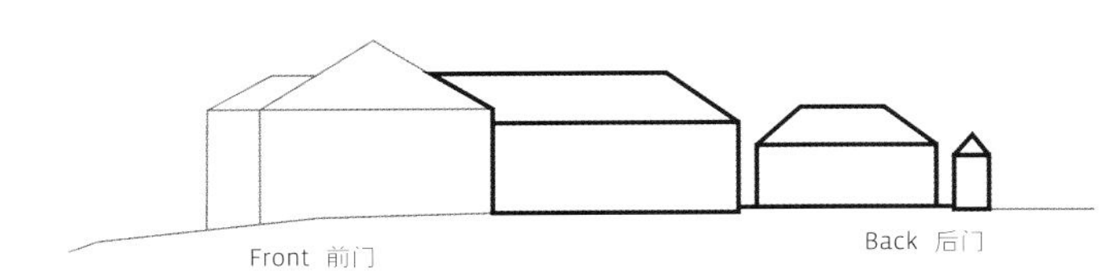

Step down in height from front to back

从前到后高度逐步下降

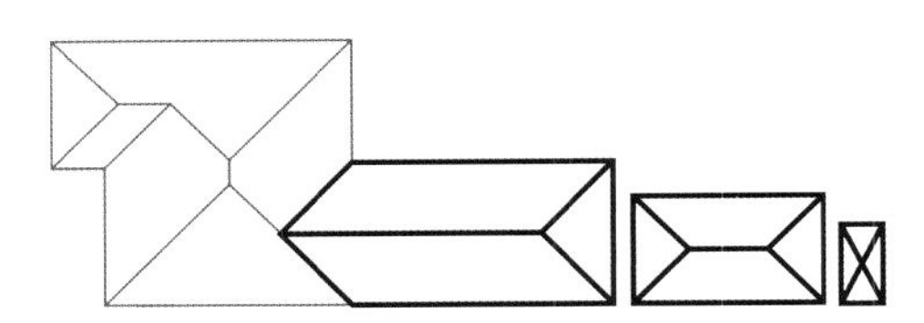

Hipped roofs form extension

斜屋脊屋顶范围大

Abstraction of roofs and extensions

抽象屋顶扩展

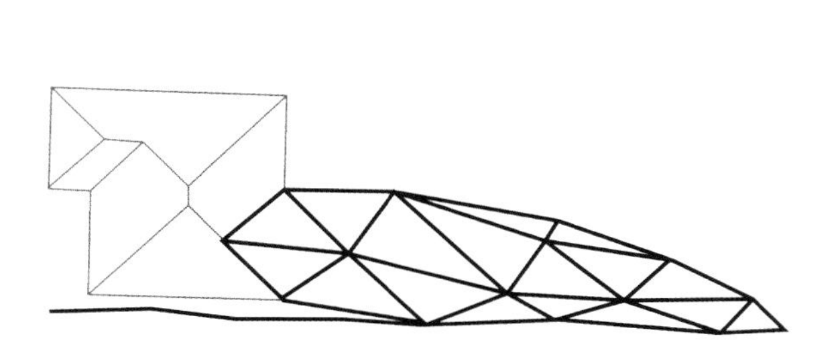

Contemporary Interpretation
现代解读

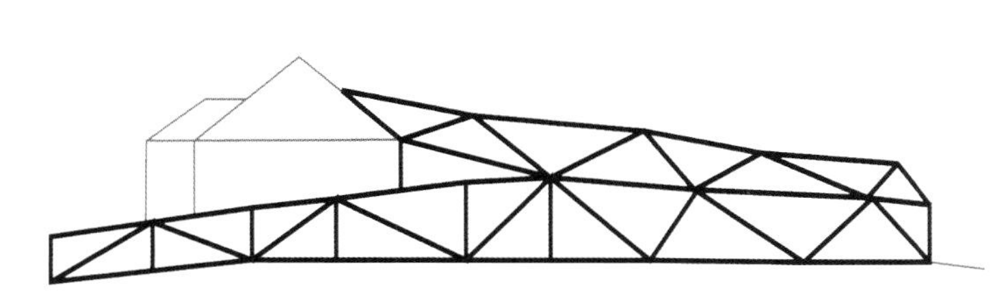

Contemporary interpretation
现代解读

这座被锈蚀铜面板覆盖的小建筑位于悉尼附近，整个建筑就像一个开敞的屋顶或棚子。这一设计理念的依据来源于该地区的一个对新建建筑的规定——新建房屋必须"佩戴"传统的斜屋顶造型。

锈蚀铜面板材料构成建筑的外墙与屋面，让建筑浑然一体。屋顶曲折起伏，达到了斜屋面的要求。其形态的设计灵感来源于四坡屋顶的三角形形式。因为基地南侧与周边建筑相邻，为了给住户创造一定的私密性，这个几何形的外壳只在北立面开窗，可以观赏到花园中种植的当地植物。

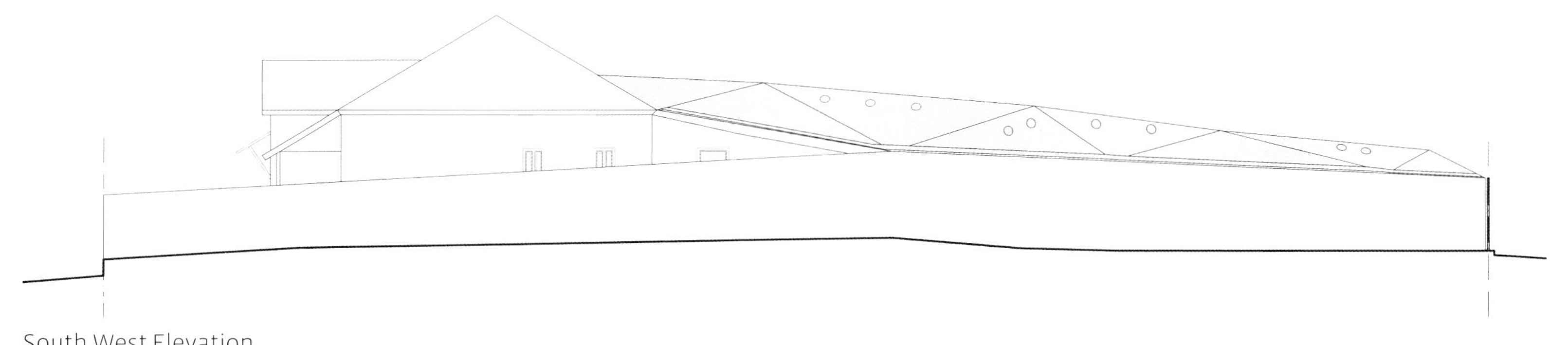

South West Elevation
西南立面

With the basic geometric forms and different combinations, lots of buildings take on a more complex form. In the project, it displays an irregular form and its plane combines simple geometric forms with polygons and in the 3-D effect, the building becomes richer in form and dynamic in peace.

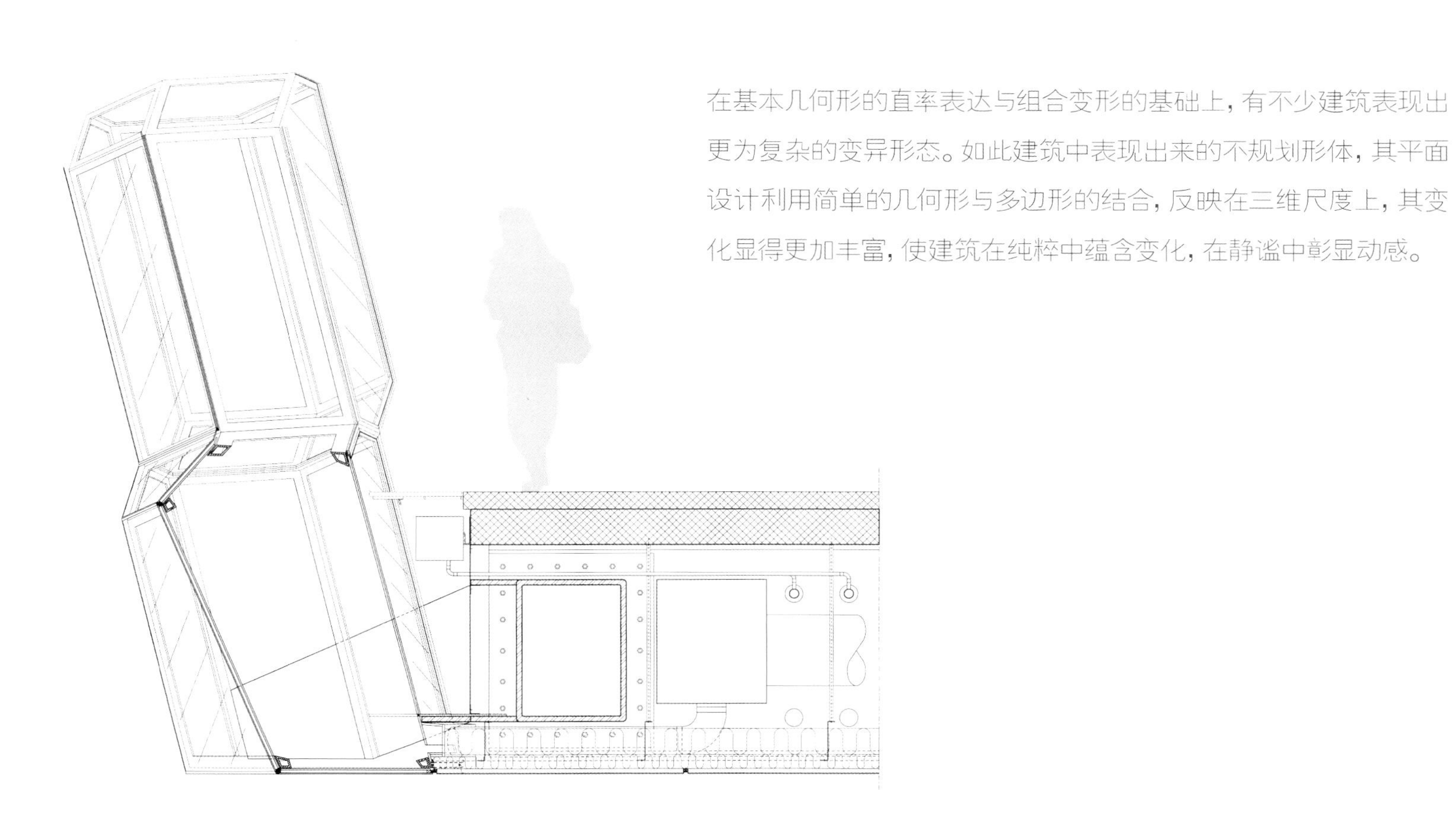

在基本几何形的直率表达与组合变形的基础上，有不少建筑表现出更为复杂的变异形态。如此建筑中表现出来的不规划形体，其平面设计利用简单的几何形与多边形的结合，反映在三维尺度上，其变化显得更加丰富，使建筑在纯粹中蕴含变化，在静谧中彰显动感。

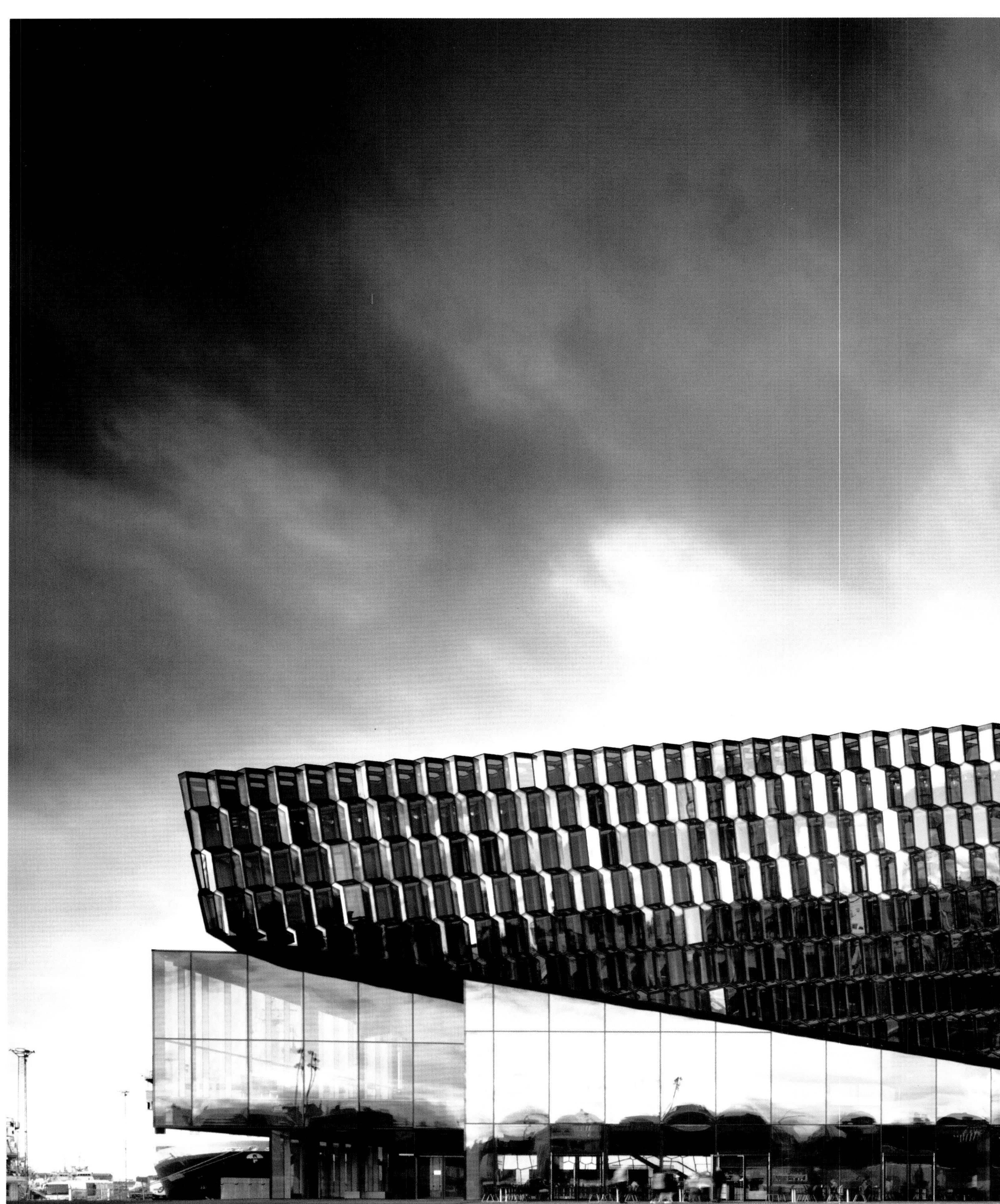

CLYFFORD STILL MUSEUM

Visitors can reach the front yard of the project after walking through trees and landscape. They can get away from the bustling city and wander into a peaceful world for relaxation. The building is clearly seen, though blocked by the trees, and contains the cast concrete wall nearby and different surfaces. The surfaces take on slim, vertical lines and form a broken, randomly placed patterns, creating rich surface texture with different shading effect in the bright sun.

游客首先通过一系列的树木和景观进入到项目的前院，在此过程中人们脱离城市的喧嚣，慢慢进入到一个安静的冥想的放松的世界。通过树木，建筑的结构依稀可见，它包含就地浇铸的建筑混凝土墙壁，还有一系列不同的表皮质地。表皮呈现出薄薄的垂直的线条，它们形成一种断裂的有机的随意的样式，营造出丰富的表皮纹理，在炽热的阳光下，呈现出不同形式的阴影效果。

Trapezoid balconies are strewn at random and with regular uneven contrast, the façade with strong rhythm matches with the bay windows, making the whole building vivid, cordial and fresh.

大面积的梯形阳台错落排列，利用有规律的凹凸对比，使建筑在立面组织上获得强烈的韵律感，它与方形窗的搭配，让建筑的平面构图更加生动、亲切而富有新鲜感。

"New york" skyscraper meets
纽约式摩天楼采用地中海式阳台

Mediterranean balcony
保养阳台

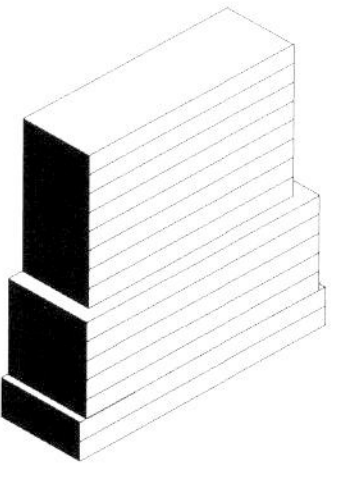

Piling Programma: Ny Setback
阶梯式后退堆叠程序

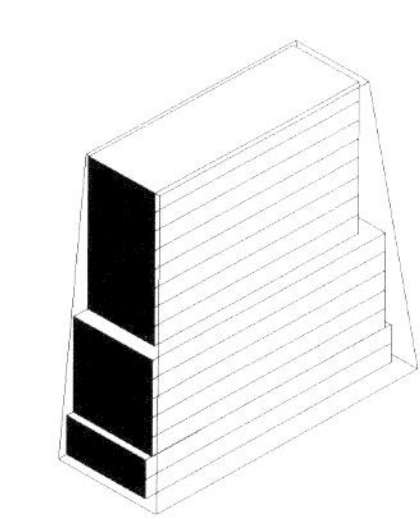

Balkonies Definine The Volume
阳台决定体量

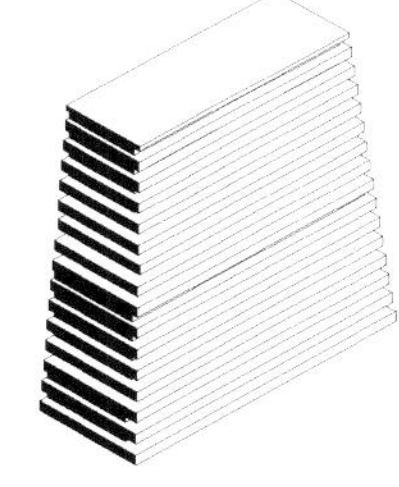

Mediterranean balcony
保养阳台

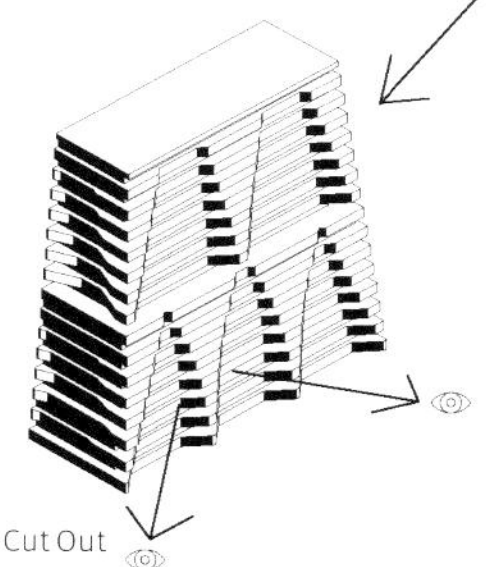

Cut Out
All Balconies Have The Same Size
裁出阳台尺寸一样

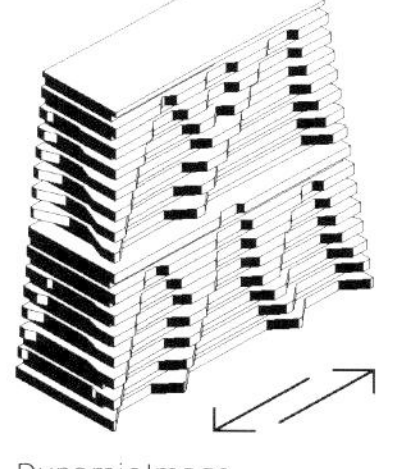

Dynamic Image
动态形象

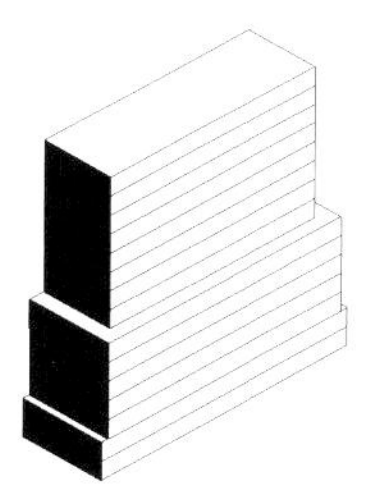

Piling Types Of Flats
平面堆叠类型

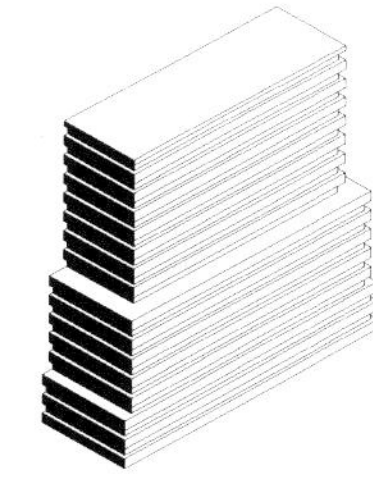

Maintainance Balconies
保养阳台

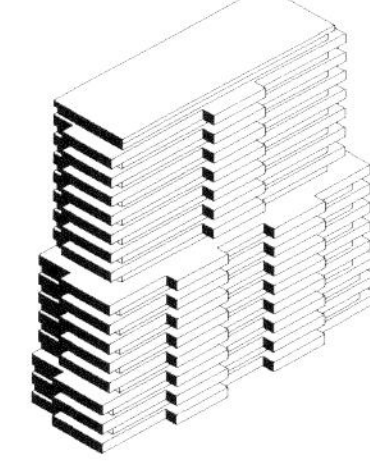

One Balcony Per Flat
每个平面都有一个阳台

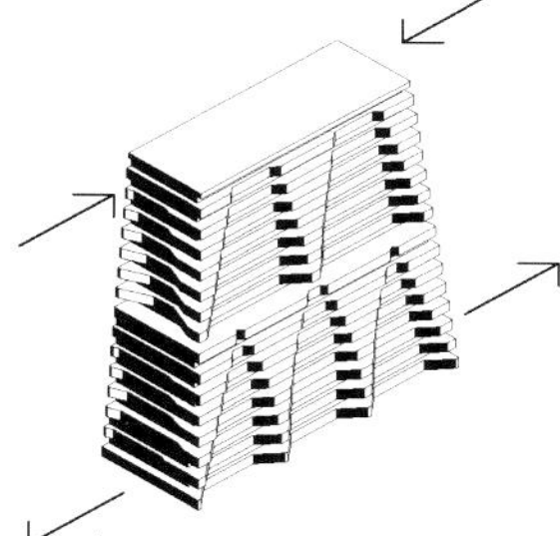

Coherent Contour
连续性轮廓

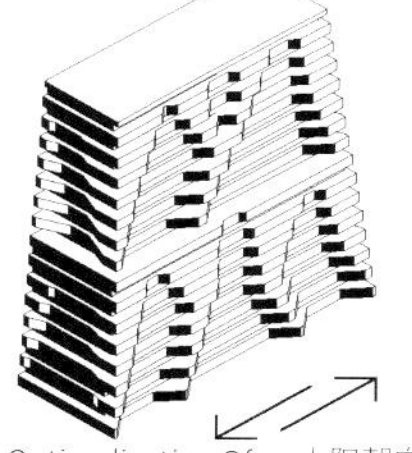

Optimalisation Of Orientation Towards Sun
太阳朝向优化

SANSIRI
VERTICAL LIVING
GALLERY sukhumvit

Nature can be felt among the buildings. Its exterior wall is stuffed with trapezoid plants and square boards in different angles yet in match. Considering the location of the project, the architect selects the local plants – Tokyo low plants as can be seen everywhere and can guard against the local extreme climate conditions. The grain of such plants softens the architectural rigidity, putting the building in the limelight.

行走在建筑中，感受自然就在身边。该建筑外墙采用梯形的植物种植和方形的木板夹层填充，横竖斜放不一，错落有致，黄绿搭配协调，给人以视觉上的满足感。考虑到项目所在的位置，建筑师选择了本土植物——东京矮生植物。这种植物在地上随处可见，可以很好地抵御当地极端的环境条件。植物的纹理也柔化了建筑的刚性，使它在周边复杂的建筑结构中脱颖而出。

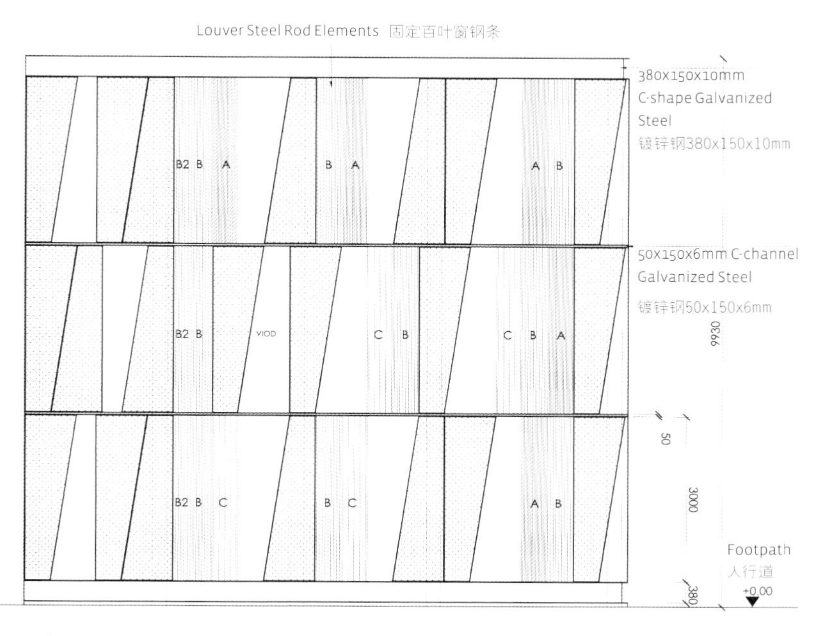

Elevation

立面

SANSIRI
VERTICAL LIVING
GALLERY

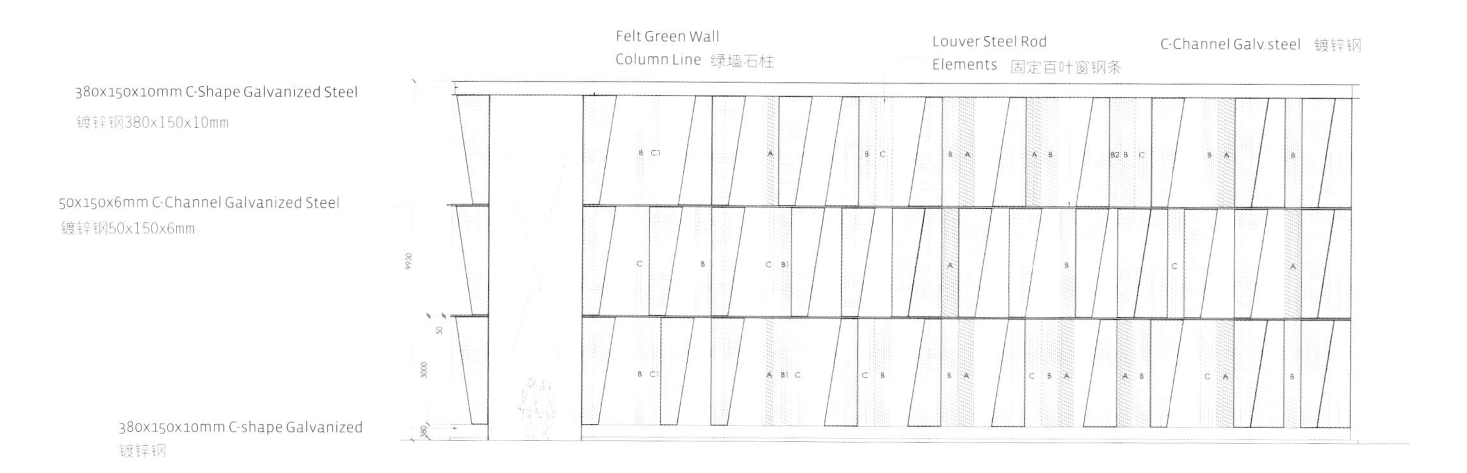
Felt Green Wall
Column Line 绿墙石柱
Louver Steel Rod
Elements 固定百叶窗钢条
C-Channel Galv.steel 镀锌钢
380x150x10mm C-Shape Galvanized Steel
镀锌钢380x150x10mm
50x150x6mm C-Channel Galvanized Steel
镀锌钢50x150x6mm
380x150x10mm C-shape Galvanized
镀锌钢

Elevation 立面

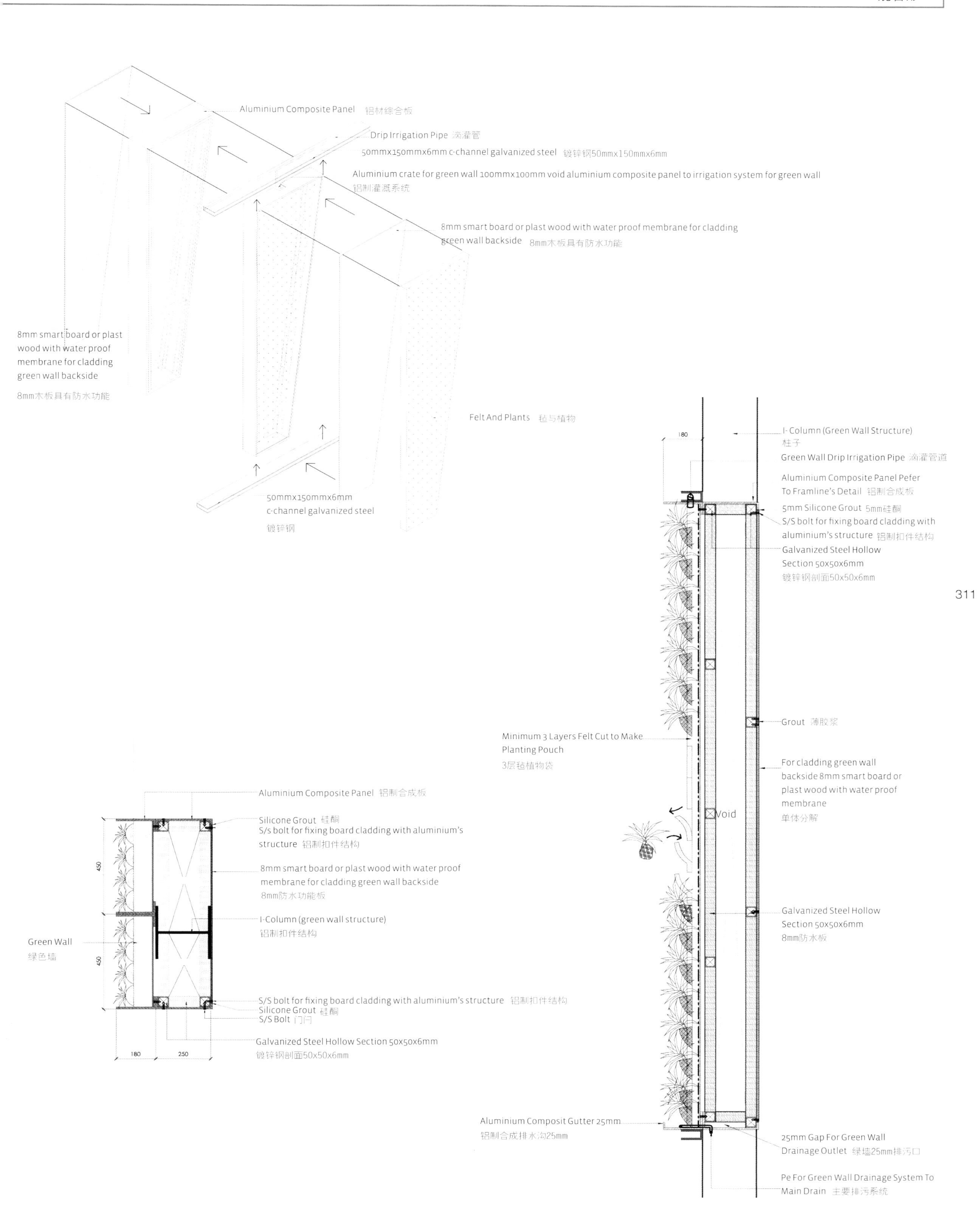
Aluminium Composite Panel 铝材综合板
Drip Irrigation Pipe 滴灌管
50mmx150mmx6mm c-channel galvanized steel 镀锌钢50mmx150mmx6mm
Aluminium crate for green wall 100mmx100mm void aluminium composite panel to irrigation system for green wall
铝制灌溉系统
8mm smart board or plast wood with water proof membrane for cladding green wall backside 8mm木板具有防水功能
8mm smart board or plast wood with water proof membrane for cladding green wall backside
8mm木板具有防水功能
Felt And Plants 毡与植物
50mmx150mmx6mm c-channel galvanized steel
镀锌钢
180
I- Column (Green Wall Structure)
柱子
Green Wall Drip Irrigation Pipe 滴灌管道
Aluminium Composite Panel Pefer To Framline's Detail 铝制合成板
5mm Silicone Grout 5mm硅酮
S/S bolt for fixing board cladding with aluminium's structure 铝制扣件结构
Galvanized Steel Hollow Section 50x50x6mm
镀锌钢剖面50x50x6mm
Grout 薄胶浆
Minimum 3 Layers Felt Cut to Make Planting Pouch
3层毡植物袋
For cladding green wall backside 8mm smart board or plast wood with water proof membrane
单体分解
Void
Galvanized Steel Hollow Section 50x50x6mm
8mm防水板
Aluminium Composit Gutter 25mm
铝制合成排水沟25mm
25mm Gap For Green Wall Drainage Outlet 绿墙25mm排污口
Pe For Green Wall Drainage System To Main Drain 主要排污系统
Aluminium Composite Panel 铝制合成板
Silicone Grout 硅酮
S/s bolt for fixing board cladding with aluminium's structure 铝制扣件结构
8mm smart board or plast wood with water proof membrane for cladding green wall backside
8mm防水功能板
I-Column (green wall structure)
铝制扣件结构
Green Wall
绿色墙
450
450
S/S bolt for fixing board cladding with aluminium's structure 铝制扣件结构
Silicone Grout 硅酮
S/S Bolt 门闩
Galvanized Steel Hollow Section 50x50x6mm
镀锌钢剖面50x50x6mm
180
250

SANSIRI
VERTICAL LIVING
GALLERY sukhumvit

EM
423
2H
05:00-22:00
EM
423
SA
424

Trapezoid roofs, square windows and marvelous curved decorations form the intriguing architectural plane, displaying the perceptual and rational beauty, namely the regular square façade, and moving curves, both of which show the beauty of life.

梯形屋顶、长方形开窗、奇妙的曲线装饰，这些构成了建筑动人的平面曲调，彰显着感性与理性的融合之美：中规中矩的方形立面，搭配着跳动活跃的曲线，严谨中显灵动，在理性的建筑表面上表达出了妙趣的生命之美。

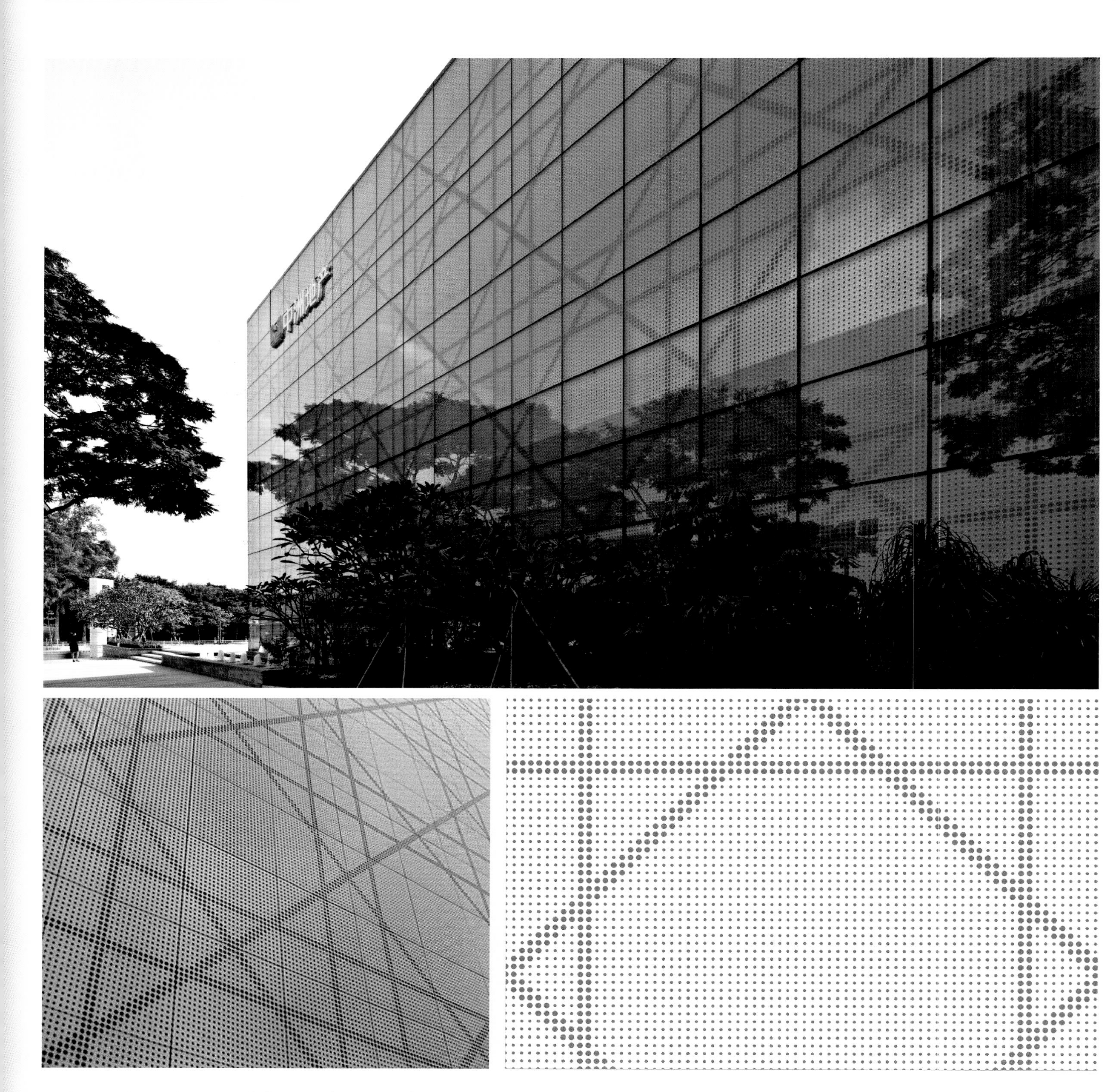

In the project, the use of slanting lines on the surface breaks the rules of square planes while the application of wire netting produces a misty effect and takes on a layered façade. Meanwhile, the originally scattered planar elements are combined, enhancing the building volume. Thus the building surface neither too heavy nor too penetrating and it is the combination of the slanting lines and reticulation, bringing a layered and exquisite design effect.

中洲·中央公园

该项目表面上斜线的运用，打破了纯粹的方形平面的规则。金属网的采用，可以产生朦胧的效果，在光影变化的时候呈现更有层次的立面表情。同时也将原本松散的平面元素统一起来，增强建筑的体量感。因此建筑的表皮介于厚重和通透之间，斜线与网状的结合，可以产生层次丰富而又精致的设计效果。

中央公园

14

6

图书在版编目（CIP）数据

变形记 : 建筑立面的衍生与突破 上册 / 香港理工国际出版社编著. — 武汉 : 华中科技大学出版社, 2013.5
ISBN 978-7-5609-8969-3

Ⅰ. ①变… Ⅱ. ①香… Ⅲ. ①建筑形式－研究 Ⅳ. ①TU-0

中国版本图书馆CIP数据核字(2013)第102770号

变形记：建筑立面的衍生与突破 上册　　香港理工国际出版社 编著

出版发行：华中科技大学出版社（中国·武汉）
地　　址：武汉市武昌珞喻路1037号（邮编:430074）
出 版 人：阮海洪

责任编辑：赵慧蕊　　责任监印：张贵君
责任校对：王晓甲　　装帧设计：香港理工国际出版社

印　　刷：利丰雅高印刷（深圳）有限公司
开　　本：965 mm×1270 mm　1/16
印　　张：20.5
字　　数：300千字
版　　次：2013年9月第1版 第1次印刷
定　　价：348.00元（USD 69.99）

投稿热线：(027)87545012　6365888@qq.com
本书若有印装质量问题，请向出版社营销中心调换
全国免费服务热线：400-6679-118 竭诚为您服务